The Practical Idealists

John and Avril Blake

The Practical Idealists

Twenty-five years of designing for industry

Lund Humphries
London

SBN 85331 237 0

First edition 1969
Published by Lund Humphries
12 Bedford Square, London WC1

Designed by
Herbert Spencer and Hansje Oorthuys

Made and printed in Great Britain
by Lund Humphries
Bradford & London

The publishers and authors wish to
thank the partners of Design Research
Unit for their help in the preparation
of this book, and in particular for
giving unfettered access to their
archives, without the inspection of
which this publication would not have
been possible. Whilst DRU have been
indefatigable in checking matters of
fact, the authors wish to make clear
that they, as authors, are entirely
responsible for any opinions expressed
and for any errors or omissions.

All the illustrations are of projects or
buildings in the design of which DRU
has been involved.

A list of photographic
acknowledgements appears on page 147

Contents

About this book

In 1942, Dunkirk and the Battle of Britain had passed into history. The steam had gone out of the German invasion of the Soviet Union and by the end of the year the Nazi army was in retreat at Stalingrad. In North Africa, Montgomery was driving Rommel back across the Libyan desert. The pattern of war was beginning to change in favour of the Allies, and in Britain, for the first time, belief in ultimate victory became no longer a matter of blind faith but a possible reality.

All the same, it seemed to us when we began work for this book, that to plan for a future brave new world at such a time was surprisingly optimistic. Bombs still fell nightly on London: yet if the Allies were going to win the war nothing, of course, must have made it more clear than the bomb-holes that unprecedented opportunities would occur, when the fighting was over, to look at old problems with a fresh eye and to make a better way of living than there had ever been before. The seeds of social and political change were beginning to germinate. Three years later they were to flower into massive support for what was taken to represent the forces of enlightenment and progress – the Labour Government of 1945.

But in 1942, disillusionment was a long way off. It was a time when lengthy working hours, separations, and all the grim realities of war made idealism, for most people, a psychological necessity. For designers and architects this was nothing new. The 'thirties had witnessed a revolution in design which was almost as dramatic and far-reaching in its consequences as the political upheavals across Europe. Awareness of the implications of an expanding industrial society had created new forms in architecture and a new approach to the design of household products. Before the war these ideas had been largely ignored or rejected by industry, but the destruction of homes, and the prospect of putting industry once again on a peace-time footing, now opened the way to the translation of pre-war ideals into post-war necessity.

Many people at this time began to think about the problem of a nation which one day would find itself without a war. It was apparent to a few that one urgent need would be to supply industry with new des gns to replace those which had been manufactured in 1939. Not only was there likely to be a huge home demand for consumer goods once

7

restrictions were lifted, but exporting, too, would become a much more important and challenging activity for many British firms than it had been before the war. Designs would need to be good once conditions became competitive, and a means would have to be found of providing practical design services which industry could draw on.

The first proposals for such a service came about as a result of a discussion between Marcus Brumwell, then chairman and managing director of Stuart's advertising agency, and his friend Herbert Read (later Sir Herbert), the distinguished design historian and critic who died last year. His book *Art and Industry*,[1] published in 1934, had come to be regarded almost as a bible by designers at the time.

Marcus Brumwell has told us about a ration-restricted dinner at his house in Surrey, in the summer of 1942, when he and Herbert Read together formulated the first ideas for a group of designers and artists who could be called on to design products for industry over a far wider range than any design office had ever done before.

Stuart's was a member of a group of smaller agencies which had banded together for mutual support at the beginning of the war. Calling themselves the Advertising Service Guild, the group shared resources, including office accommodation for bombed-out members, and together supported and later took over Mass Observation, one of the earliest market research organizations. Brumwell and Read discussed their proposals with Cecil D. Notley who was then head of Notley Advertising, another member of the Guild. As a result of these discussions, Notley invited Milner Gray to set down on paper a detailed plan for the kind of design service they felt would be needed. Gray sent his ideas to Notley in a letter dated October 1942, and the plan he put forward laid the foundations for the events we shall describe in this book.

Gray, at the time, was head of the exhibition design department of the Ministry of Information. He wrote to Notley: 'The purpose of the formation of a group such as is envisaged is to make available immediately a design service equipped to advise on all problems of design, and to form a nucleus group which, through contacts

[1] *Art and Industry*
Faber & Faber 1934

established during the war period with specialist designers and experts in all appropriate fields, would be in a position at the end of the war to expand and to undertake the wider services which may then be demanded of it. The final aim is to present a service so complete that it could undertake any design case which might confront the State, Municipal Authorities, Industry or Commerce.'

Milner Gray's note went on to describe the organization of his proposed design group in some detail. It was clearly his suggestion for a three-part structure embracing the functions of design, research, and administration which anticipated the title which the new design group was eventually given – Design Research Unit.

The Unit was set up on 1 January 1943, and Herbert Read was its first manager and sole member of staff. Last year, in 1968, DRU reached its silver jubilee, and we were commissioned by the publishers to write a book to commemorate the event. Our assignment, however, was not to produce a conventional history of one particular organization, but to use the jubilee occasion to survey in broader terms the course of design progress since the war.

Within this context, we have used DRU as an example to show how the major trends and changes have affected the practice of design for industry and, in turn, how design practice has itself influenced the course of events. This is why, although the work of other designers is referred to in the text, all the illustrations are of DRU's designers and their work. We do not wish to suggest, however, that the Unit is supreme among the design offices which have grown up in Britain during the past twenty-five years; indeed a similar book could have been written around one of several other design organizations.

But there are, nevertheless, special reasons why DRU can serve as an exemplar for the design profession as a whole. It is probably the oldest, largest, and best-known industrial design office in Europe; the people responsible for its growth have made, as individuals, profound contributions to the ideas which have enabled industrial design to be accepted throughout the world; its origins, as we shall see, spring from what was almost certainly the earliest group design consultancy in Britain; and, in recent years, its clients have included some of Britain's largest industrial companies.

In writing this book, however, we found that our assignment – to use a particular example to illustrate the more general theme – proved to be unexpectedly difficult. The problem became increasingly like that of looking at a middle-distance object through bi-focal lenses. But in the end we felt that the kind of dialogue that emerged; between events seen in panorama and in close-up, had been helpful in relating the evolution of design theory to the real-world challenge of providing an effective design service for industry.

John and Avril Blake
Maplehurst, 1969

Chapter 1 **Prelude**

Origins of industrial design

No discussion of recent developments in industrial design can ignore their historical background, for events over at least the past century and a half have created problems to which the solutions are only now beginning to emerge.

Designers who are irritated by too much theorizing, rightly point out that people have always designed products for use either by themselves or by others, and that what designers are doing today is really no different from a creative activity which has been practised since the beginning of human history. But there is a difference, of course; a difference which many historians and writers on art and design have noted was most closely associated with the industrial revolution. Some writers on design history have given the impression that at some point in Victoria's reign there was a clean break between an age of one-off, hand-made artefacts, and an age of the machine in which products were suddenly turned out in vast quantities for a mass market. This is an exaggeration, for the industrial revolution was only part of a continuing change which was comparatively slow up to the beginning of the nineteenth century, but has subsequently become both rapid and diverse. Hand work in industry has, in fact, persisted in varying degrees right through to the present day, with tools which are sometimes not so very much more sophisticated than those used in the eighteenth century or even before. At the other extreme, it can be argued that quantity production has existed since the first printing press, or even since the templates of the earliest civilizations.

The significance for design of the industrial revolution was an acceleration of development as a result of mechanical power, producing a complex interaction of social, economic and technical change. The balance which had existed in earlier and slower times became upset. This was not merely a simple imbalance between design and technology but between one form of design and another and between various aspects of technology in a whole variety of permutations.

Nikolaus Pevsner in *Pioneers of Modern Design*[2] vividly describes the situation existing around the turn of the eighteenth century which resulted from the profound changes that were taking place. 'The immediate consequence of this precipitous development was a

[2] *Pioneers of Modern Design*
Faber & Faber 1936
Penguin Books, revised edition, 1960

11

sudden increase in production, demanding more and more hands, and so leading to an equally fast increase in population. Towns grew up with horrifying rapidity, new markets had to be satisfied, an ever bigger production was demanded, and inventiveness was stimulated anew . . .

'In the midst of this breathless race, no time was left to refine all those innumerable innovations which swamped producer and consumer. With the extinction of the medieval craftsman, the shape and appearance of all products were left to the uneducated manufacturer. Designers of some standing had not penetrated into industry, artists kept aloof and the workman had no say in artistic matters.'

It was this situation which led to a growing concern for the 'artistic' quality of manufactured products. The concern was expressed by the Government, because it could see that an expanding industrial economy would depend on industry's ability to satisfy the tastes of consumers at home and abroad; and it was expressed by writers and artists, and the educated classes generally, who were concerned more with the effect of quantity-produced products on the cultural well-being of society.

Official action to remedy the shortcomings was taken in 1836 by the Ewart Committee, which was set up to enquire into 'the best means of extending a knowledge of the arts and of the principles of design among the people, especially the manufacturing population', leading to the establishment of the National School of Design and to the Victoria and Albert Museum, then called The Museum of Ornamental Art.

Concern was also expressed in publications such as *The Journal of Design and Manufactures* founded by Henry Cole, and by critical letters to the Press expressing horror at some of the products shown in the Great Exhibition of 1851.

The significance of all this interest in design, however, is that it was directed predominantly towards one area of production: the kinds of product which had been made for domestic use for hundreds of years – things like textiles, furniture, ornamental metalware, pottery and glass. The form and decoration of such products had evolved over the years by a slow process of modification and refinement by the craftsmen who made them. It was a process which everybody understood, and a tradition in which everybody could participate. When the new production methods of the industrial revolution took over from hand methods, it was inevitable that an attempt should be made not only to imitate the results of the craft tradition but, because of the new skills available, to 'improve' on them. The problem was that the 'improvements', as Pevsner has described, were being made by a new generation of manufacturers who lacked the craftsman's disciplines arising out of direct contact with the materials he worked with and the customers he served. The products of the new age could be seen directly against the norm of the craft tradition, and for the discriminating spectators, the comparison was not a very happy one.

Their concern, however, wholly ignored another area of product manufacture: the development not of the products which were made by machines, but of the machines which made the products. Strangely, there appear to be few books which deal with the history of engineering design, although there are plenty concerned with the technologies on which the process of design in the engineering industries depends. Apart from passing references to the railways, design historians have largely ignored mechanical engineering altogether, although they have, of course, paid a good deal of attention to the civil engineering structures of people like Telford, Brunel, or Paxton, and to the attempt to make the alien forms of machines more acceptable by applying what today sometimes appears to be inappropriate decoration.

The study of these aspects of engineering is part of the first-year course of every industrial design student, but the purpose of such studies is to enlarge the student's understanding of the influence which the engineering industries have had on the 'artistic' qualities of products. It does not add up to anything like a philosophy of engineering design or, to be more precise, a total design philosophy capable of embracing all forms of engineering design as well as industrial design. The attitude underlying these studies is understandable, however, for the design activity concerned with engineering products has evolved over a hundred years or more with practically no contact with the design for industrial production of the kinds of product previously designed and made by craftsmen.

The industrial revolution thus saw the creation of two entirely separate design streams which, virtually, have remained divided right up to the present day. The first of these streams has been concerned predominantly with the artistic content of products and will be referred to in this book as 'the art-oriented stream'; the second is concerned with the application of logical principles and quantifiable information, and will be referred to as 'the science-oriented stream'.

No doubt designers and engineers will rush to point out that a great deal of engineering design is 'artistic' in the sense that decisions are based on an intuitive preference for one solution as against another, or that the design of any pot or piece of furniture must clearly be based on logical principles. This, of course, is true, but the distinction between art- and science-oriented designing, while clearly an over-simplification, is broadly correct and explains the confusion and difficulty in finding common ground which occurs whenever design is discussed in 'mixed company'.

There is not only a difference of approach between the art- and science-oriented streams, there is also a difference of objective. Art-oriented design is concerned with the achievement of desirable formal properties; science-oriented design with the achievement of desirable performance characteristics. Increasingly both are concerned with meeting these objectives within economic limits.

In the context of this brief look at the origins of design in an industrial age, the term 'industrial design' requires further explanation because, to some extent, it spans both the art- and science-oriented streams. It comes directly from the nineteenth-century concept of 'industrial art' which was originally used to describe industrially-produced ornament, as distinct from hand-produced ornament. Industrial design carried this a stage further since it referred not merely to the ornament applied to products, but to the design of the product as a whole, and similarly was used to distinguish this process from craft design. As such, it tended to become associated primarily with the kinds of product which had been made by hand, and these were mostly in the consumer goods field. This association has persisted until the present day and is evident in most of the products which are now on show in The Design Centre.

The activity of designing products which have craft origins can normally be accomplished by one person alone, in consultation with the manufacturer's production and marketing experts. Since the designer is concerned with the whole product, rather than a part of it, his objectives must include both those of the art- and the science-oriented streams. A tea-pot, for example, must have desirable formal properties and meet obvious performance requirements. Its designer could therefore be described as a 'complete designer', but he will know, nevertheless, that it is the artistic qualities of his tea-pot which will give positive satisfaction to the user, while the effectiveness of its performance will be taken for granted.

The achievement of desired performance in this kind of product does not put excessive demands on the technical capabilities of the designer. A different situation, however, has come about with the development of the science-oriented design stream. In the early days of mechanical and civil engineering, the technologies involved were comparatively simple, and the knowledge required was capable of being embraced by one person in just the same way as it was for the craft-based products. Development was by a process of trial and error and it was only later that laws and principles became established. The engineers responsible for the early steam locomotives, cast iron bridges and factory machines achieved a unity of design because they themselves were capable of controlling all aspects of the product they were designing. Subsequently, however, the technologies required in the design of engineering products developed so rapidly that, in most cases today, no one man is capable of acquiring all the skills needed for a single job, nor does he have time to do all the work himself. Increasingly, engineers have had to specialize, so that a man skilled in the structural design of buildings, for example, would need to do a lot of homework before he could

tackle a hydro-mechanical braking system for a car, and would probably
be completely baffled by the comparatively simple circuit design of a
television set. The idea of a 'complete designer' for the engineering
industries has been replaced by the need for design teams in which the
unity that was characteristic of early engineering design is much more
difficult to achieve.

In the nineteenth century, the philosophies of Ruskin and Morris
succeeded only in widening the divergence between the two major
design streams and to exacerbate what possibility there was of any
understanding between them. More important, however, is that the
Bauhaus, which stands in people's minds for the total rejection of the
Ruskin/Morris attitudes, did very little, at least in its early stages, to
take design thinking outside the banks of the art-oriented stream. In
her book on the Bauhaus,[3] Gillian Naylor quotes Gropius's
proclamation at the opening of the school:

'Let us create a new *guild of craftsmen,* without the class distinctions
which raise an arrogant barrier between craftsman and artist. Together,
let us conceive and create the new building of the future, which will
embrace architecture and sculpture and painting in one unity and which
will one day rise towards Heaven from the hands of a million workers,
like the crystal symbol of a new faith.'

Much of the theorizing about design in the years between the two
world wars was directly in line with these ideas. In Britain, interest was
still dominated by a concern for the artistic qualities of craft-based,
consumer products, and it is significant that in 1930 the name of the
first professional organization representing industrial designers was the
Society of Industrial Artists, and that this name remained in being
until 1965 when the words 'and Designers' were added, not without
opposition by some members.

This concept of injecting an understanding of art into industry
dominated most of the official thinking about design throughout the
'thirties. When the Gorrell Committee was set up by the Board of
Trade in 1931, to consider the desirability of establishing permanent
and temporary exhibitions of 'good design', it published its report the

[3] *The Bauhaus*
Studio Vista 1968

following year under the title *Art and Industry*. Three years later, as a
result of the committee's recommendations, a new Council was set up
with the term 'Art and Industry' again in its title, and throughout its
life it was concerned almost exclusively with products like textiles,
jewellery, pottery, and ornamental metalware.

Even Herbert Read, who had pioneered so much of the new thinking
about design, was concerned with the problem of whether or not
'the machine' was capable of producing a work of art. Two aspects of
this concern are relevant: the first is that 'the machine' seemed to be
regarded in a curious way as some sort of abstract phenomenon,
existing remotely, mysteriously, and rather menacingly in the
background, like the factory in Fritz Lang's film *Metropolis*. (That
factories still employed people to paint patterns on pottery by hand,
for example, or made furniture by comparatively simple tools and jigs,
did not apparently alter this strange concept of the machine as a
Juggernaut.) The second aspect is that the machine itself or, to be more
accurate, the extremely diverse products of the engineering industry,
were either ignored as objects for which design in any sense was
important or were idolized in a curiously romantic way as a symbol of
the new scientific age.

In reality the engineers were quietly working out their own solutions
with little awareness of the artistic controversy which surrounded their
efforts. If the result of their work was considered to be visually good,
then this was a justification of their belief that form was an outward
expression of function. If it were not, then either the critics were
wrong or beauty was unimportant.

By the 'thirties, however, manufacturers of certain mass-produced
engineering goods, like cars, domestic appliances, and office equipment,
began to see that the engineering process did not automatically result
in the kind of beauty that the customer wanted. In America,
particularly, all forms of sales promotion were needed to stimulate
demand, and designers like Raymond Loewy and Walter Dorwin
Teague were soon able to show manufacturers that they were losing
business because of the muddled appearance of their products.
Industrial designers, skilled in the achievement of the kind of formal

visual quality which had been developed in products for the craft-based industries, could provide the missing skill and tidy up the mess which the engineers had left.

Loewy himself was an engineer, but there were others who were not, and sometimes the most they succeeded in doing was not so very different in intention from the nineteenth-century idea of industrial art. They earned the tag of 'chrome-strip artists' and, while this term is not in use today, the impression died hard in the minds of engineers who are still suspicious of industrial designers encroaching on what they consider to be their own rightful preserves. Not understanding the peculiar origins of the term 'industrial design', they prefer the word 'styling' which, in spite of protestations to the contrary, describes fairly accurately the main contribution to engineering design which the industrial design profession has made until the last few years.

The beginning of design practice

One of the activities of the Council for Art and Industry was to publish in 1937 a report on *Design and the Designer in Industry*. The report saw that one of the difficulties in the way of more rapid improvements in design standards was the poor training and lowly status of designers employed on the staff of consumer product manufacturers. What training they received was either given by one of the 150 or so art schools which had sprung up throughout the country during the previous century, or by apprenticeship within the firm. The result was hardly likely to provide a breed of designers capable of raising the artistic sights of their employers. The report stressed the importance in this situation of consultant designers, saying of them: 'The fresh mind of these craftsmen and artists and their experience in other and freer fields may bring to industry a revelation and an inspiration not otherwise obtainable'.

Strangely enough, however, the beginnings of consultant design practice in this country stemmed more from the advertising explosion than from the artistic inadequacies of consumer goods. Printed paper of all kinds was being disgorged in ever-increasing quantities from agencies and from company advertising departments. Posters, Press

advertisements, packaging, and display material provided a new system
of communication between producers and consumers, and a massive
flow of designers was needed to keep the system going. Much of the
work produced was of a poor quality, since the directors of many
agencies, like the majority of the consumer goods manufacturers,
believed that the interests of commercialism could not be equated with
the aesthetic ideals of those whom they looked on as 'do gooders'.

But there were important signs of change. From its beginning in 1915,
the Design and Industries Association had paid a good deal of attention
to graphic design and the real possibility of art, in this two-dimensional
sense, serving the needs of industry. The 'thirties had witnessed a
unique renaissance in British painting and sculpture, and artists such
as Henry Moore, Ben Nicholson, Barbara Hepworth, Graham
Sutherland, and others were beginning to establish a reputation
throughout the world. Earlier, Edward Johnston's lettering and
McKnight Kauffer's posters for the Underground had led to the idea of
the new tube stations as galleries of art for the masses, with the poster
as a medium for public enlightenment. And in spite of Sir Kenneth
Clark's statement to the DIA in 1935 in the first issue of *Trend*
magazine, that 'on the whole the public prefer ugly objects', a few
leading agencies were searching for new talent among artists and
designers who could prove the falsehood of this supposition.

They found them in a few design groups which were formed in direct
response to this challenge. The first of these was started by Milner
Gray and Charles and Henry Bassett in 1922. It was called Bassett-Gray
and described itself, in an early leaflet as 'the distributing organization
of a body of artists who design for industrial and commercial purposes'.
The artists included Graham Sutherland and other painters; also an
increasing number of designers, five of whom, in 1935, joined Milner
Gray in a reorganized group called Industrial Design Partnership. One
of these designers was Misha Black, who had joined Bassett-Gray the
year before, and this began his association with Milner Gray which has
lasted ever since. Other members of the partnership included Gray's
brother, Thomas; Jesse Collins; James de Holden Stone; and Walter
Landauer. Graham Sutherland, Rowland Hilder, the designer Clive
Gardiner and others continued to work with the group as associate

members – setting a prototype pattern for Design Research Unit, which was to come into existence eight years later.

In the early years, Milner Gray won widespread recognition for his meticulous draughtsmanship, which was soon reflected in his approach to typography and graphic design generally. His interests, however, were always wider than the personal responsibilities of his own design practice, and he took an active part in the promotion of good design. He was a founder member of the Society of Industrial Artists in 1930, became its president for the first time in 1943, and was elected again in 1968, the first person to become president for a second term.

His interest in design education also stemmed from the pre-war period, when he was principal of the Sir John Cass College of Art and a visiting lecturer at a number of other schools. Over the years, it would be no exaggeration to say that at one time or another he has been a member, chairman or president of almost every committee or organization concerned with graphic or industrial design, including international bodies like the Alliance Graphique Internationale.

His often apparent nervousness, which conceals an irreverent humour, could hardly be more unlike Misha Black, whose calm, measured arguments have dominated countless conference platforms and boardroom tables. Misha Black's ability to grasp the essentials of a situation has been a major factor in the development both of the pre-war partnership and, later, of DRU. Like Milner Gray, he has been associated with many bodies concerned with design and has been president of the Society of Industrial Artists and of the International Council of Societies of Industrial Design. Outside the office, his major commitment since the war has been his work as Professor of Industrial Design (Engineering) at the Royal College of Art.

The setting up of Industrial Design Partnership suggested a shift towards product design and this reflected an expansion into such fields as glass, pottery and plastics which had already begun before the Bassett-Gray group changed its title. The emphasis, however, continued to be on graphics, packaging, and exhibitions. IDP was anxious to show that the steady flow of old-fashioned, tasteless and muddled graphic design work still being produced not only made few pretensions to art, but was also of doubtful commercial value. High standards of design, they argued, far from conflicting with commercial needs, would actually satisfy them more effectively. Milner Gray's packs and labels for firms like Kodak and Hartley persuaded more and more firms that good graphic design was not just an extra intended as a boost to prestige, but a skill of central importance in modern marketing.

Industry was slower, however, to see how the same principles of logic and common sense used in graphic design could be applied to the design of their products. It was easy enough to see the connexion between the kind of training provided by the art schools and the two-dimensional work of the graphic designer, but not how this kind of experience could be adapted to work in three dimensions. It is possible, too, that many designers themselves were unsure of their ground when faced with the unfamiliar problems of construction and assembly. Nevertheless, there were some notable achievements in consumer product design in the 'thirties, especially in some of the new technology-based industries. The radio manufacturers E. K. Cole took on leading designers, among them Serge Chermayeff, Wells Coates, and Misha Black, to produce distinctive new shapes for radio cabinets using plastics materials; while their rivals, Murphy, collaborated with Gordon and Dick Russell to make an equally distinguished range of designs in wood. The Gas, Light and Coke Co. also employed Misha Black to design a heater which, in its restrained classical lines, looks as modern today as when it was introduced in 1938.

But these were exceptions. Describing the more usual attitude of industry to the industrial designer at this time, in a paper he gave to the Royal Society of Arts in 1949, Milner Gray referred to resentment against 'the pretensions of such outside "experts" by manufacturers who were unable to accept that anyone could know more about their products and markets than the manufacturer himself or the people he employed'. Yet it was certainly the ambition not only of IDP, but of other design offices, to work increasingly in the consumer products field.

This ambition was understandable because it was evident that many of the products on sale to the public were ugly, pretentious, and often incompetent in design. It could never be satisfactory for a group of people, inspired as they were by a reforming zeal, to spend their lives devising better methods of selling incompetently designed products. Nor was an isolated product of good design in a manufacturer's catalogue of horrors doing more than scratch the surface of what the new profession felt was needed.

Thus, even at this early stage, the idea of a comprehensive design policy for a manufacturer began to take shape. IDP saw the immense possibilities of a design programme which began with the design of a company's products and extended through its packaging, advertising, and display to the design of showrooms and exhibition stands. But there was seldom the opportunity to carry out such schemes, and this was the measure of the gap between the ideal and the reality of design for industry at that time. Although the ideas had been formulated in principle, they were to materialize only after a further generation of struggle.

Where IDP did make substantial inroads into the field of three-dimensional design was in exhibitions and interiors. Exhibitions were seen to be an important development in extending the activities of the advertising industry. Trade fairs became increasingly significant as a method of promoting the sale of new products both at home and abroad. Later, during the war years and after, they came to be regarded as a propaganda medium, a method of selling ideas rather than products, which led, as we shall see later, to the development of the narrative and often waggish style that has persisted up to the present day.

One early example of exhibition design, by Misha Black in 1935, was a portable exhibition stand for use at agricultural shows. The main constructional feature of this was its demountability, yet it seemed a 'solid' white rectangular building characteristic of pre-war 'modern' architectural styles. There was certainly no hint of the spidery, space-frame exhibition structures which were to become so popular later on. But his 'Kardomah' Café and shop-front in Piccadilly,

designed a year later, was prophetic of things to come. Had it been
retained and renovated, instead of being replaced a few years ago by
a tawdry design imitating the fashionable styles of ten years earlier,
it would today have been exactly in keeping with the current interest
in the best aspects of design in the 'thirties.

IDP was wound up in 1940, just after the beginning of the second
world war. In the same year Milner Gray began his war-time job at the
Ministry of Information, where he was joined by Misha Black and two
other early members of DRU, Kenneth Bayes and Dorothy Goslett.

Kenneth Bayes, an architect, had joined IDP in 1937. His arrival
had marked an important development in the partnership because it
was one of the earliest examples in Britain of graphic design, industrial
design, and architecture being directly associated in one practice. This
experiment proved so exciting and rewarding that it provided a basic
pattern for Design Research Unit. Recently, Bayes has been specializing
in architectural criteria for buildings for maladjusted and retarded
children. His report on 'The therapeutic effect of environment on
emotionally disturbed and mentally subnormal children' is the result
of a Kaufmann International Design Award – a series of annual awards
which, until recently, were given for achievement in various aspects
of design by the Edgar J. Kaufmann Foundation. This extra-mural
study has led to consultative work in the United States and to requests
to write and to lecture on the subject. Kenneth Bayes now hopes to
instigate the formation of an information and research centre in Britain
dealing with environmental design for all types of handicapped people.
He has also made a special study of the work of Rudolf Steiner, the
designer of the Goetheanum building near Basel, whose architectural
ideas have a positive bearing on therapeutic environment. In his
report, Bayes refers to Steiner's belief that 'architectural forms work
so deeply into the human being that they affect the moral life of man,
and that this will apply increasingly in the future in such a way that
architecture will be able to affect behaviour defects such as lying
and stealing.'

Dorothy Goslett, who was a campaigns officer in the exhibitions
division at the Ministry, became, after the war, DRU's business

manager. Office administration is usually considered to be a back-room job, but Dorothy Goslett has widened the context of her work by using her experience to write a guide for all embryo design offices. In 1961 she published a book on *Professional Practice for Designers*[4] which set out all her accumulated knowledge on the subject.

'Designers . . .', she wrote, 'have to deal directly with business men, who are seldom likely to be sentimental about creative work. The design services they buy have eventually to sell and sell well. The relationship between designer and client is therefore strictly a business one. The designer's codes of professional conduct are the foundation on which this relationship is built and they protect both sides of it. But the rest of the structure depends to a great extent on the designer: not only on his talent but on the degree of efficiency with which he runs his office.'

Dorothy Goslett's book has contributed a good deal to a general understanding that, while most designers have a generous proportion of idealism in their make-up, they must also be practical business men if they are to turn their ideals into reality without becoming bankrupt in the process.

In the next two chapters we shall look at the changing events and ideas which have influenced the evolution of design for industry since the war. We shall also see how these affected Design Research Unit and how the Unit developed as an organization providing a service, and expecting an appropriate reward.

[4] *Professional Practice for Designers*
Batsford 1961

Chapter 2   **Austerity**

Establishing a group practice

The five years immediately after the war have been described as 'the age of austerity', when the arrival of peace was disconcertingly accompanied by even greater privation than had been known in the war. It was a period of contrasts, when the queues and shortages, the 'Work or Want' slogans shouted from the hoardings, were matched by a Utopian faith in social security, national health, equality of educational opportunity, and the marriage of technology and art in the planning of new towns as well as in the paraphernalia of everyday life.

'It is difficult to recall a time', wrote Michael Sissons and Philip French in the introduction to a recent paperback[5], '. . . when TV was only a metropolitan toy, ball point pens a source of wonder, and long-playing records a transatlantic rumour. . . . The great social experiment that was being conducted gave rise to a sense of crusading idealism and to virtually all a feeling of involvement in national affairs.' Those concerned with industrial design and architecture had a vision of a world containing nothing but mass-produced objects of simple beauty, and cities of tall buildings in steel and glass. Such an environment seemed so obviously right to the people who promoted it that it would have been inconceivable to them that a few years later John Betjeman would be campaigning for the preservation of Victorian buildings, or later still that young people who had experienced neither war nor austerity would delight in pastiches of any period from Robert Adam to Odeon.

The emphasis in the post-war years was on a break with the past, on a new approach which avoided the imitation of previous styles and created a new design tradition arising naturally from the production processes of the twentieth century. The period also saw a great expansion in the number of design consultancies. They were established, as the pre-war Council for Art and Industry had suggested, to bring a wider range of design experience to the problems of individual companies. Many firms employed no industrial designers on their staffs, and those that did were sometimes persuaded that a consultant's more objective view could provide staff designers with a fresh approach.

Some consultancies consisted of individuals practising on their own

[5] *Age of Austerity 1945–1951*
Hodder and Stoughton 1963
Penguin Books 1964

or with a few assistants. Others combined together in groups offering
a variety of skills. Design Research Unit was an example of the second
approach and, in an early leaflet explaining the aims of the Unit, a
single sentence argues the case for group practice:

'Like every aspect of modern industry', it said, 'design should be a
co-operative activity, and the function of DRU is to focus on every
project it undertakes the combined knowledge and experience of
several creative minds since it believes that only by pooling the talents
of a team of designers is it possible to offer a service capable of
meeting every demand from the wide and varied field of present-day
industry.'

In 1942, when Herbert Read, Marcus Brumwell, and Milner Gray had
formulated their first ideas for a group service, there was little
immediate scope for it. A solution had therefore to be found which
would, nevertheless, establish a valid service to those manufacturers
in a position, even in war time, to commission design work.

Herbert Read agreed to become director of the Unit and installed
himself in a small office in Kingsway from which he set out to
introduce to potential clients the new service the Unit offered. In many
ways Herbert Read was the most unlikely of people to find himself in
charge of an operation intended to convince a highly sceptical industry
that a new-fangled activity called industrial design had any relevance
to its interests. A poet, literary critic, philosopher, and acknowledged
authority on design, he was the gentlest of men, with a manner so shy
and quiet that when he spoke it was often difficult to hear what he
said. But he knew most of the leading artists and designers and was
liked and respected by them. The contacts with industry, on the other
hand, were known to Marcus Brumwell, Cecil Notley, and other
members of the Advertising Service Guild, including Everett Jones
and Rupert Casson. Their task was to bring the two sides together, to
create a nucleus of designers and to persuade industry to employ them.
With this in mind, Herbert Read co-opted DRU's first team of
associates, which included Milner Gray and Misha Black, the architects
Frederick Gibberd and Sadie Speight, the structural engineer Felix
Samuely, and the designer Norbert Dutton, all of whom continued to

run their own practices, except for Black, Gray and Dutton who were then still employed at the Ministry of Information.

Slowly the Unit got under way, but by the end of the first year the problem of publicizing the service, and of organizing the few commissions that were received, required more time than Herbert Read was able to give in the two days a week he had been able to spare for it. Early in 1944, DRU advertised for a full-time manager. Among those who applied was Bernard Hollowood, an economist by training, who was then a teacher at Loughborough College, and subsequently became widely known as editor of *Punch,* from which he retired in January 1969. He had been deeply interested in design, less from an idealistic point of view than from the standpoint of an economist who was well aware how important design would become after the war if Britain was to achieve the 75 % increase in the volume of exports which Sir Stafford Cripps had announced would be essential. He got the job, but stayed for only three months as he was then faced with the counter-attraction of being offered the assistant editorship of *The Economist*.

Hollowood has recalled his excitement at the prospect of working with Herbert Read, whom he had admired from his writing but had not met. 'It was like God descending', he said. 'It was always an effort for him to become a man of the world and, although he had other business interests, he always longed to be back in the peaceful surroundings of his Yorkshire home.'

He has remembered travelling to the Midlands with Read on a visit to a number of firms. They were not always welcomed very enthusiastically, but Herbert Read persuaded one motor manufacturer to employ the sculptor Naum Gabo to design a post-war car. Gabo's design was an aerodynamic shape incorporating an all-Perspex canopy over the passenger compartment. It was in fact an early hard-top concept but, although Read was enthusiastic about it, the manufacturer found it too revolutionary.

Bernard Hollowood's work, when he went to *The Economist*, was taken over by Bill Vaughan, who had been a colleague of Milner Gray's at the Ministry, and who stayed on with DRU until 1946, when the

founder partners were released from their war-time jobs and the
Unit came to life as a more fully active organization.

The year before this, Gray had designed an exhibition called *Design
at Home* which had been organized by the Council for the
Encouragement of Music and the Arts (later the Arts Council). It
opened on VE day, 8 May 1945, and contained examples of utility
furniture in room settings which firmly established the new concern
for a sense of order, with carefully aligned surfaces and an airy
spaciousness, in keeping with the 'brave new world' ideals.

The development of utility furniture during the war had been a key
factor in the popular acceptance of an austere style of design which
was quite different from the quasi-traditional designs characteristic of
the majority of furniture produced in the 'thirties. It was a style which
had come about from necessity – from the shortage of materials and
labour – yet the very economy of its design and production was
directly in line with the ideas of functional simplicity which stemmed
from the Bauhaus.

Gordon (later Sir Gordon) Russell was at that time chairman of the
Board of Trade's Design Panel responsible for the utility furniture
scheme. He writes of it in his autobiography,[6] '. . . there wasn't enough
timber for bulbous legs or enough labour for even the cheapest
carving, and straightforward common-sense lines were efficient and
economical. It was an acid test and . . . must have been a bit of a shock
that a type of design which had been pioneered for years by a small
minority – whilst the trade looked on and laughed – should prove its
mettle in a national emergency, but so it was, to the amusement of
some and to the amazement of others.'

While inevitably, as the furniture industry got back on a peace-time
footing, there was a reaction against the utility outlook, the fundamental
common sense of the best war-time furniture had been assimilated
both by the trade and by the public, and had a small but noticeable
effect on the design of other products for the home.

[6] *Designer's Trade*
Allen & Unwin 1968

A major spur at this time to the new generation of post-war designers

and architects was the setting up, in December 1944, of the Council
of Industrial Design. After more than a century of Government
committees and reports, some real action to improve design standards
had been taken at last. A report on *The Visual Arts,* prepared by
Political and Economic Planning in 1946, emphasized that since
'. . . every effort to improve design in the past has failed for lack of
continuity, it is to be hoped that the new Council will receive adequate
support over the period of years which it must take to establish itself'.

The need for time became increasingly obvious as the work of the
Council got under way. But the results have been sufficiently
rewarding for successive Governments to maintain support for the
Council with enough money to ensure the continuity which the PEP
report realized was essential.

Although from its earliest days the Council understood that industrial
design had a part to play in the heavy engineering industries, its policy
was to concentrate first on consumer goods, where the effect of any
design improvements it could make would be most readily apparent.
In the event it took another twenty years before it could get to grips
with industrial design in engineering, and to make the first realistic
but tentative plans for joining the art- and science-oriented streams.

During the war years little had been done to prepare new designs for
consumer goods, either to meet the market at home or the new markets
which would have to be found abroad. In many cases designs which
had appeared in the late 'thirties were used again after the war. Writing
of this period, Gordon Russell, who was director of the Council from
1947 to 1959, said 'industry as a whole was complacently unaware
that anything was wrong with the goods it produced. Undeniably it was
true that they could all be sold easily at the moment, but the seller's
market would pass, competition would greatly increase in world
markets and then the suitability and look of the goods – their quality –
would become as important as price and delivery. When that time
came it would be no use trying to set up a council in a hurry, for an
immense educational job had to be done – starting with reorganization
of the whole method of training industrial designers, persuasion of
industry to employ them, the creation of public interest in better

designs and the consequent stimulation of a demand at home which would in time have an influence on export trade.'

Going to work on an egg

One of the first important activities of the Council was to provide this public stimulus in the form of a major national exhibition of industrial design, called *Britain Can Make It*. At the start, the new Council had little real knowledge about what standards of design were being achieved by those industries which had been able to get back to peace-time production. To cover themselves against possible failure to find sufficient acceptable designs, the CoID briefed the exhibition designers to produce a display that would stand, if it had to, by itself. The exhibition opened in September 1946, by which time diligent searching had produced a surprisingly good selection of products. But the briefing resulted in an exhibition that was far more than just a collection of manufactured goods. It also set out to explain what was involved in the design process, and to show the need in industry for the special skills which the industrial designer could provide.

This part of the exhibition was given to DRU to design and organize, and proved to be the Unit's most important commission until that time. It was designed by Misha Black and, using an egg-cup as an example, explained how even the simplest mass-produced product creates a multitude of problems which it is the designer's job to solve. The effect of production processes on design was shown in sections on pottery, glass, plastics, textiles, and metalware. But that processes must be the servant and not the master of the designer was conveyed by a photograph of a plastics moulding press which announced, on a card, 'Cannot Design – Can Only Make', while another label hung on the machine apologetically proclaimed 'No Brains'. To prove the point a display of some sixty egg-cups, each one different in shape, illustrated the large area of freedom still left to the designer. It was a highly educational approach and it set a pattern for narrative displays which has been characteristic of British exhibitions ever since. It also revealed a British fondness for coating educational pills with a whimsy which today seems a trifle corny, but at the time clearly delighted the crowds who queued – nearly half a million strong – to shuffle through the exhibition. For the majority of these visitors, however, the products shown elsewhere in the exhibition would merely whet their appetite, for the pleasure of buying was largely to be denied them. Decorated china, sumptuously fitted kitchens, and new domestic products of all kinds were to remain in short supply for another four or five years, most of them being destined for 'export only'.

The description of the industrial design process given in the Egg exhibit was part of a much wider interest at that time in explaining the function of the new activity of industrial design to an industry which at best was unfamiliar and willing to learn, but otherwise was highly sceptical. In 1946, DRU's partners and associates, with others not connected with the Unit, contributed individual chapters to a book on *The Practice of Design*[7] for which Herbert Read wrote an introduction. A small booklet produced by DRU about the same time, and circulated widely to industry, summed up the main theme of the larger book in more general terms. To some extent the ideas which it expressed seemed hardly to have changed since before the war. It said, for example, that 'Industrial design is the intelligent, practical and skilled association of art with industry'. And, later, in describing the qualities that the industrial designer could bring to the service of industry: 'First and foremost, the industrial designer is a fully qualified technician. He (or she) may be an architect, a scientist, an engineer, a fellow of the Society of Industrial Artists, but whatever the technicalities of his training he is basically an artist – one who has been trained to know good proportions, clean lines, well blended or contrasted colours, and who can demonstrate this on paper with brush and pigment.'

But later it went on to emphasize the practical skills which the industrial designer could provide, and a section on design research is particularly significant in explaining the seriousness of purpose which DRU wanted to put forward as the special quality of post-war design services. 'Successful design', it said, 'is based on data, not on inspiration alone. . . . What is the existing or potential market, its extent and type? What competition exists? What are the comparative advantages of the principal competitive product? What possible advantages have not been developed? What are the opportunities for enlarging and developing the market to other fields?'

[7] *The Practice of Design*
Lund Humphries 1946

That research is an essential preliminary to design, was a refrain which was repeated at every opportunity in the early years, and was certainly needed to dispel the popular idea that industrial design was the last-minute 'tarting up' of a product of which the essential form had already been settled. The ability to express this need in convincing terms rapidly developed in the years which followed. In his paper to the Royal Society of Arts in 1949, which we referred to in the previous chapter, Milner Gray expanded on this philosophy in terms which are as relevant today as they were at that time:

'In Great Britain', he said, 'during the second world war, the production of armaments and essential goods imposed on industry necessary economies in skill, time, material, and effort. In certain fields, such as the development of aircraft production, and the technique of packing war-like stores, to quote only two examples, the encouragement of design closely integrated with production became a sheer necessity for survival. And so in the nineteen forties the attitude of the more informed propagandists for the better design of machine-made goods has developed a bias towards the needs of production. *Design for selling* has given way to *design for making*. From the cosy gatherings of the converted and the high ideals of the highbrows of the nineteen twenties, industrial design comes down to earth.'

It was design for production, he went on to say, which brought into focus the new need for detailed planning and research. 'The separation of the manufacturer from his market, implicit in a highly mechanized society, compels him to re-establish contact with it through market research. . . . You cannot make 100,000 copies of an article which nobody wants – at least you are not likely to be asked to do so more than once.'

But fact-finding about the market was not the only reason he gave for research. Facts about production equipment and methods, and about distribution and handling, were all essential preliminaries to design, for only after all the information had been properly assembled, analysed and digested could the design work itself begin. 'But in the long run', Milner Gray concluded, 'it is his aesthetic faculty which differentiates the designer from the mere technician.'

Although it had been the intention of DRU at the beginning to offer a complete service of research and design along the lines set out by Gray in his paper, the idea was found to be too ambitious to be realizable in practice. How could any design organization hope to encompass research in depth for all the materials and technologies, the production methods and markets, and for all the industries which it would encounter in the course of its work? For one thing, industry was not prepared to pay for this amount of work; for another, new research organizations, both private and state supported, were beginning to be set up which were far more capable than a design office of carrying out the types of research which were required. At the same time DRU soon found itself with more design work than it could easily cope with. So, like its export research organization, which had been rather optimistically set up in 1946, DRU's intention to establish research services soon became submerged in the tide of events.

The need for reorganization

Extra staff had to be found to deal with the growing number of design commissions that were coming into the office and several young assistants were taken on. At about the same time, the 'first wave' associates were beginning to find that their own practices were expanding fast and that they could do less and less work for DRU. Until then the Unit had been huddled into three rooms in Bedford Square. 'In one large room', Dorothy Goslett remembers, 'there were Misha, Milner, myself, Austin Frazer, Jo Revill, the plan chest, the boardroom table, and almost everything else, and one telephone between the lot of us.'

Reorganization was clearly needed and this became possible when Marcus Brumwell bought up all the shares from the Advertising Service Guild and became the Unit's sole owner, establishing a holding company to control its finances. This change made for a much tighter direction of financial policy and DRU moved again, this time to its own offices in Park Street.

A new form of associateship was introduced, for a group of designers who, after war service, wanted to set up in practice for themselves but had insufficient capital to do so. DRU offered them office space, services, and a financial pool from which they could draw each month an agreed sum, which was later set off against their total earnings. In return, the associates gave priority to DRU work, over and above their other commissions. Their final income from the Unit was worked out on the basis of an agreed proportion of the fees for each job; a system which left them, at the end of each year, either with a lump sum due to them or, more rarely, with a deficit to be cleared.

Such a system worked well for DRU because it enabled them to use designers of a higher calibre than they could have afforded to employ on a full-time basis. It also suited the associates because it provided them with a base for their operations, a degree of financial security and sufficient freedom to enable them to build up their own practices.

The structure of the office at this time thus consisted of Marcus Brumwell, the financial controller, who delegated executive responsibility to a partner of Stuart Advertising, H. C. Timewell; Herbert Read, director; three founder partners, Gray, Black, and Bayes; the business manager, Dorothy Goslett; several associate members, who included Robert Gutmann and Ronald Ingles; and, increasingly, a number of younger salaried designers and architects such as John Diamond, Clifford Hatts, Herbert Spencer, and George Williams.

The Unit was organized on strictly democratic lines. There was no 'boss' who handed out the work and told other people what to do. It was essentially a group operation among friends in which each job was discussed on the basis of equality among everyone concerned and the work carried out by whoever was thought best qualified to tackle it. To maintain this sense of shared purpose and democratic exchange of ideas, a regular six-weekly discussion meeting among all design staff and some clerical personnel was established, and has remained sacrosanct as an institution throughout the entire history of DRU.

At these meetings work on current jobs is presented and explained by the designers concerned, and everybody has an opportunity to comment or criticize. Although they are informal, the meetings are conducted by a chairman – one of the partners – and were usually attended by Sir Herbert Read until his death. A guest critic is

A meeting of the DRU Board of associated designers in March 1948. *Back row, left to right,* are Marcus Brumwell, Robert Gutmann, Brian Peake, and George Williams. *Front row, left to right,* are Sadie Speight, Frederick Gibberd, Milner Gray, Dorothy Goslett, Bronek Katz, Herbert Read, Misha Black and Clive Entwistle.

sometimes invited, someone from industry or the trade press, partly to introduce an objective view from outside the Unit, and partly to give the meetings a secondary use as a public relations medium.

But perhaps more important to the work of the office than either of these purposes has been the educational aspect of the discussions for everybody involved. Few punches are pulled when work is being criticized, and designers soon become skilled in defending their work on a logical basis. A designer whose ideas will be challenged by his colleagues will make sure that each step in the design process has a reasoned explanation – in itself a discipline which can only lead to better standards. Moreover, if his design has been 'thought through' in this manner, the designer will be in a far stronger position when presenting his work to the client across the boardroom table than would otherwise be the case. He will be used to the need for clear explanation, and will have gained confidence from the experience of speaking to an audience. At the same time, as a critic, his perception will have been sharpened and his understanding widened by having learned that even among designers there can be an extraordinary diversity of attitudes. The younger designers, particularly, are encouraged to take part in the discussions, not only as part of their own training but also to teach, since the senior partners frankly admit that they look to the younger members of the Unit to keep the rest aware of new attitudes and ideas.

Planning the Festival

DRU expanded rapidly during the late 'forties, but in spite of the emphasis on product design in the Egg exhibit at *Britain Can Make It*, comparatively few commissions for product design work came into the office. What did seem to impress potential clients who saw the exhibition was the skill with which the Egg exhibit had been designed. This led to further commissions for exhibitions, including *Darkness into Daylight*, a display of lighting equipment sponsored by the Electric Lamp Manufacturers' Association; *Design at Work*, an exhibition at the Royal Academy, organized by the CoID and the Royal Society of Arts, of work by members of the Faculty of Royal Designers for Industry; and a growing number of stands at trade shows such as the British Industries Fair.

Even before *Britain Can Make It*, however, discussions had begun about a far more important proposal – an international exhibition to mark the centenary of the Great Exhibition of 1851. The original public proposal for this was made by the late Sir Gerald Barry (still Mr Barry at that time) in an open letter to Sir Stafford Cripps published in 1945 in the *News Chronicle*, of which Barry was then editor. A year later, a drawing was published in *The Ambassador* showing an earlier proposal by Misha Black for using the derelict South Bank of the Thames for such an international fair. But it was to be another four years before all the proposals and counter proposals, and a seeming multitude of committees, had arrived at a final plan for a scaled-down national exhibition designed to demonstrate to the world Britain's confidence in her future as an industrial nation. *The Festival of Britain* became the greatest single stimulus to experiment and creativity in architecture and design that the country has ever experienced. Never before had so many designers come together to work for a common aim, and seldom had they been given such an opportunity to explore new design ideas in a mammoth expression of twentieth-century idealism.

By 1947 a design group had been appointed to plan the exhibition in outline. This was led by Sir Hugh Casson and included James Holland, James Gardner, Ralph Tubbs, and Misha Black. By 1949 it was finally agreed to use the south bank site after practically every other possibility had been considered and rejected. Soon after this a master plan was approved and detailed design work began.

The site was divided in two by the arches of the Hungerford Bridge, and this feature, in conjunction with the irregular shape of the site as a whole, led the panel to decide on an informal arrangement of buildings, instead of the classical avenues arranged in a geometric pattern, which had been characteristic of the 1937 Paris exhibition and other international fairs. This freedom of design, with spaces flowing into each other in an apparently haphazard manner, was to have a lasting influence. It stemmed from an essentially romantic national character which in previous ages had produced the picturesque landscapes of Capability Brown, and the work of the great topographical painters of the late eighteenth century. It is also reflected in the British abhorrence for the grid-iron cities of America and, in later years, led to an entirely new approach to the design of the urban scene in both new towns and old.

The Hungerford Bridge also provided a natural division of the site into 'upstream' and 'downstream' areas for organizational purposes. Sir Hugh Casson, as architect; James Gardner, as designer; and R. T. James, as engineer, were responsible for co-ordinating the downstream area, while the upstream area was supervised in the same way by Misha Black, Ralph Tubbs, James Holland, and Ralph Freeman.

For DRU, like so many other design offices at the time, designing for the Festival was an overwhelming preoccupation for two or three years before the opening on the third of May 1951. Under Misha Black's overall direction, Kenneth Bayes led a team of nine designers who prepared all the displays for the inside of Ralph Tubb's Dome of Discovery; and Alexander Gibson, who joined the Unit in 1948, designed the Regatta Restaurant, the most important architectural job that had been carried out by DRU at that time. Gibson also co-operated with the War Office and the LCC's chief engineer on a Bailey bridge for pedestrians across the Thames, beside the Hungerford Bridge. On the graphic design side, Milner Gray directed another team of designers concerned with the development of the complete sign-posting scheme for the exhibition.

But there were plenty of problems to contend with. The first five months of 1951 were the wettest for many years. A series of strikes held up the work and one, at the end of January, brought everything to a standstill for a fortnight. The Bailey bridge collapsed into the river and had to be fished out and started again. On the night of the third of May, after King George VI had declared from the steps of St Paul's that the Festival was officially open, there were still exhibits waiting to be set up, including some in the Dome of Discovery. The construction workers knocked off at five o'clock that evening, leaving the designers responsible for the inside of the Dome to hump the remaining exhibits into position themselves, working through most of the night. In this way the Dome was finished before the King led a party of invited guests on a private tour of inspection the following morning. It was a leaden day. By lunch-time rain had again set in and DRU's designers went to their own Regatta Restaurant for the first full meal they had seemed to have had for a long time. Through its clear glass walls they watched for the arrival of an excited public, let in for the first time after the official reception was over. When the public came, splashing through the puddles, far from being excited, they looked wet and depressed. The designers shared their depression. After all the hard work, they seemed to be faced only with anti-climax and exhaustion.

But this mood did not last long. The designers were to have their triumph within two days, and it was to last as long as the memory of the Festival.

'London's Gayest Night for Years', reported the *Sunday Express* on the sixth of May 1951. 'Fairyland at Midnight.' 'Traffic chaos.' 'Packed five deep on Waterloo Bridge they had a grandstand view of the brilliantly lit South Bank.' 'Crowds milled round the Dome of Discovery, the Skylon and various pavilions, looking and wondering.'

Raymond Mortimer, in *The Sunday Times*, rather reluctantly admitted at the end of a lukewarm report: 'It is all such fun and the feeling of enterprise and enthusiasm is infectious . . . Flying staircases, huge ships, seaside resorts in miniature, tractors ploughing the air, cafés that look like cranes or parachute-schools, dissolving views of geological changes, everywhere there is something fresh and fanciful. . . . We may be bankrupt of money; we may, like the rest of the world, be on the road to bankruptcy in the fine arts; we are not – this at any rate is manifest – we are not bankrupt in inventiveness or zest.'

The idealism of the previous decade seemed to have been embodied in tangible form. A new self-confidence was in the air and nothing, it seemed, could prevent the upward swing into a new age of prosperity. Though nobody would claim that the Festival was in itself responsible for the new wave of optimism that began to spread across Britain, it so exactly mirrored the changing climate of the period that it is now seen as the turning point from austerity to affluence.

Festival in retrospect

Looked at in retrospect, how has the Festival stood up to the test of time? More specifically, what were the contributions it made to architecture and design, and what influence have these contributions had on subsequent history? There can be no doubt that the greatest impact was made by the architecture and by the specially designed features of the townscape. It is difficult now to realize that London in 1951 was a city which could boast hardly a single major building in the modern manner. The GPO was still building quasi-Georgian post offices. No tower blocks pierced the city's skyline. Architectural students were fed on a diet of Corbusier's *Maison Suisse* and Tecton's pre-war Highpoint flats. The box-like prefabs which covered many of the bomb-sites had been Britain's only serious contribution to architecture since the war. In this context, the Festival Hall, the only building that was to remain a permanent monument to all the glitter and excitement of the Festival, seemed like a soaring, airy palace to those used to promenading uncomfortably in the echoing spaces of the Albert Hall.

But the Festival Hall was by no means typical of the architecture at the exhibition. In fact the chief characteristic of the Festival architecture was that almost every pavilion was, in itself, an individual and experimental design. The massive upturned saucer of the Dome, with its gloomy interior, was in complete contrast to the spidery, glass-walled structure of the Transport Pavilion immediately next door; the severely elegant Regatta Restaurant and the whimsically idiosyncratic Lion and Unicorn Pavilion were at opposite ends of the spectrum. The massive laminated arches of the main entry building showed a delight in structural experiment that was repeated in the Skylon, but the two structures were none the less totally different in concept. The one thing that tied them all together was their delight in newness for its own sake, and the faith in the ability of technology to solve all problems.

Most of the people who contributed to a symposium in *Design* magazine ten years after the Festival, agreed that there had been no recognizable 'Festival style', although they thought that there had certainly been a Festival spirit. Misha Black, who was one of the contributors, described how this spirit resulted from a positive and coherent control arising from convictions commonly shared by the architects and designers concerned, and therefore willingly accepted by them.

This control applied as much to the products that went inside the buildings as it did to the buildings themselves and to the landscaping and display design. The task of selecting the exhibits was the specific responsibility of the Council of Industrial Design, but both the challenge and the resources available to meet it were much greater than they had been for *Britain Can Make It*. It was to be a testing operation for Gordon Russell, and he has recalled that Gerald Barry's original intention was to have a separate pavilion of industrial design, but that he persuaded Barry that every product in the exhibition ought to be subjected to the same rigorous and independent examination. 'Otherwise', he has written, 'we might find clumsy W.C.s and basins in the public lavatories, all sorts of lettering in the different pavilions, the cheapest park chairs for seating – in fact a general impression that lack of co-ordination had produced the particular brand of chaos in which Britain tends to specialize.'

The decision to go ahead on these lines was of fundamental importance
to the subsequent work of the CoID and to design in general. The
selection of 10,000 or so products needed for the exhibition was an
immense job, and to carry it out the Council recruited a team of
about twenty industrial officers, each one of whom was responsible
for an industry or a group of related industries. A photographic
record, divided into seventy categories, was devised as a basic
source of reference for the architects and display designers.
Such a comprehensive collection had never before been made, and the
result proved so interesting that it was put on display on the
South Bank as *Design Review*. After the Festival it was retained as
a permanent index of well-designed British products, and its
maintenance, by a continuous process of selection and revision, has
remained a central activity of the CoID ever since. Without it, the
establishment of The Design Centre in 1956 could hardly have been
contemplated.

It is tempting to speculate about the effect on public appreciation of
design if the comprehensive approach to the selection of products for
the Festival had been continued in other fields, such as public
purchasing or trade exhibitions overseas. But if the idea was ever
considered seriously, it was quickly dismissed as probably being
impracticable and certainly undesirable. As if to expunge any taint of
authoritarianism which might have been left over from the Council's
role in the *Festival of Britain*, the CoID has gone out of its way ever
since to emphasize its belief that only persuasion and not compulsion
will be effective in the long run.

Where the Festival could be said to have produced a distinctive and
recognizable style was in the entirely new interest which was created
in the spaces surrounding and linking the individual pavilions. This
interest is significant not so much for the spectacular fountains and
sculptures, some of which were outstanding in their quality of
inventiveness, but for the more modest treatments of the paved
surfaces and planting arrangements. Everywhere people walked, they
were conscious of the designer's hand. The use of cobbles, bollards,
pools, and plant troughs created a prototype for a new way of treating
urban spaces. Today many architects and planners, concerned with

new towns and urban redevelopment, have similarly exploited opportunities to enrich the townscape.

A distinctive style was also apparent in the displays, and in the kind of decoration that was used. There was a particular interest in models of molecular structures and the patterns of life-forms revealed through the microscope. Referring to the influence which these had had on 'the philistine manufacturers', John Murray, economist, architect, and journalist, who also contributed to the anniversary symposium in *Design* magazine, wrote: '. . . these Frankensteins of our unconscious making have leered out at us from miles of misbegotten textiles and acres of jazzy carpets during the past decade'.

Similar criticisms were made by graphic designers and typographers of the decorated Victorian type-faces, and their derivatives, which had been a particular feature of the Festival, for these were considered to have had a disastrous influence during the years immediately following the 1951 exhibition both on print design and on the use of lettering on buildings.

Such criticisms, however, were being made a decade later, when informed taste was advocating a style in architecture, typography, and product design of a severity which had not been experienced since the Bauhaus. The reaction among the *avant-garde* to the gaiety, colour, and decoration of the Festival was swift and profound. The 'new brutalism' was born almost before the gates of the South Bank were finally closed. The popularity of Festival decoration was its downfall. But from the longer viewpoint of almost a second decade away, the mood has changed again sufficiently to allow the Festival to be seen in a wider perspective.

The reaction undoubtedly exaggerated the harm which may have resulted. The flower children of the present generation might well have felt as comfortably at home among the coloured knobs and flags of the South Bank, or the baroque whimsy of Battersea Funfair, as in the Kings Road or Carnaby Street. This is not merely an expression of a further swing of the pendulum. It represents the beginnings of a more realistic philosophy of design which we shall discuss later in more detail.

Chapter 3 **Affluence**

Signs of commercial prosperity

The American magazine *Look,* in a special issue published in 1959, described the decade which was then drawing to a close as 'the fabulous 'fifties'. 'America', it said, 'enters an age of everyday elegance.'

The people of the States had never before experienced such economic prosperity, and many Europeans looked enviously at the technological marvels that were coming out of American factories, or were being presented in exhibitions as a way of life which was only just around the corner. It was a period of dream kitchens, where the housewife was shown at her control centre keeping an eye on the family through a television screen, pushing buttons which selected cameras in different parts of the house, provided automatic meals, or instructed the robot floor-cleaning device to go into action. Cars became longer and sleeker, and created a new symbology of status and power. It was a world in which happiness could be found in possessions, and the possessions were smarter, bigger, and more modern in America than anywhere else.

This at least was the impression among those in Britain whose horizons in the early 'fifties were being modestly widened by an unfamiliar ability to buy as many eggs as they wanted. Festival year had marked the end of rationing, but durable consumer goods were slow to appear in sufficient quantities to provide the public with much choice.

When *House and Garden* magazine, in January 1952, proclaimed in bold headlines 'New Year's Resolutions: Be Greedy! Be Lazy! Be Inquisitive! Be Over-fond! Be Self-indulgent! Be Extravagant!' it seemed like blasphemy to a generation brought up to the idea that National Savings were the most sacred of sacred cows. It was a dramatic break from the penny-pinching attitudes that had been a mark of the domestic economy until then. The beginnings of the affluent way of life had arrived in Britain, and were later to have a profound effect on design thinking among manufacturers, consumers, and designers.

But at the time when the magazines were pointing to a new prosperity in Britain, the designers who had helped to create this image were themselves far from prosperous. The Festival had tempted DRU, like other design offices, to an over-expenditure on time and effort which sapped its resources. The administrative complexity of what had been the biggest co-ordinated exhibition design project of all time had also complicated the work of the Unit, which found itself dealing with numerous Government departments and other bodies. For three years, planning and design work for the Festival had taken priority over most other work and, as opening day drew nearer, everything had to be sacrificed to the overriding need to get the job finished on time. The staff of DRU had risen to forty to cope with this volume of work, but the sudden hiatus after the opening left the Unit dangerously weak from a financial point of view.

Fortunately, this did not last long. In 1952 DRU was asked to take part in a scheme which introduced an entirely new sense of affluence, at least to the business world if not to the general public – the interiors of a new building in Bond Street designed by Michael Rosenauer for Times Life. Sir Hugh Casson and Misha Black again worked together to co-ordinate the design, and they brought in some of Britain's leading designers to work with them on specific areas – Leonard Manasseh, Neville Condor, Robin Day, Neville Ward, and others. Paintings and sculptures by Henry Moore, Ben Nicholson, Maurice Lambert, and Geoffrey Clarke were specially commissioned, and the Council of Industrial Design was sufficiently impressed by the lead this American company was giving to British industry that it published one of the first special issues of *Design* magazine to review the project. In an introduction by the architect Lionel Brett, the interiors were hailed as the beginning of a new vernacular in British design 'and will undoubtedly be held to represent the advanced taste of the 'fifties as surely as the BBC interiors represent that of the 'thirties'.

After Times Life, there was an increasing demand for the design of interiors and, in the following years, commissions of this kind formed a substantial part of DRU's work, gradually replacing for the Unit's architects the design of exhibitions as their main activity. Some of this work was concerned with a variety of comparatively small showrooms and shops, but a major job, for the interiors of BOAC's new headquarters at London Airport, was completed in 1955. This was almost in total contrast to the luxury of Time-Life, for the chief requirement was the layout of the great engineering hall in which the working environment had to promote the maximum operational efficiency. Special attention had to be given to the design of work benches, component trays, and other equipment, and the exacting demands of the project led to long consultations with BOAC's staff, work study trials, and a good deal of re-designing. The results were enthusiastically received by those who were accustomed to using the old equipment, and there was an equal enthusiasm for the colour, lighting, signposting, and final over-all appearance of the hall.

On these early interior design projects, administrative procedures were used which DRU first evolved for the Time-Life building, and which have continued to be used ever since. All equipment selected by the designers is coded to ensure that everything arrives at the right time and place. Ordering, delivery, and checking are simplified to the point where supervision can be left to comparatively junior administrative staff, thus freeing the designers for more creative work. At the same time the budgets are broken down into small units, allowing a tight control over costs to be maintained throughout the job.

This sort of efficiency, quite as much as their design skills, won for DRU a continuing series of major interior design commissions. In the last decade these have included work in the P & O passenger liner *Oriana*; the Royal Garden Hotel; *The Times* new building; West London Air Terminal; the CIS building in Manchester; new offices for the Alliance Building Society in Brighton; Britannic House, the new headquarters of British Petroleum; and many others. The last four of these examples were office buildings for big national companies and show that, as the Council of Industrial Design had hoped, the lead given by Times Life had at last been followed by British industry.

Although DRU's architects were directly involved in these interior design projects, the success of the Regatta Restaurant at the *Festival of Britain* had encouraged the Unit to hope for increasing opportunities to design buildings as a whole. In the event, there were disappointingly few commissions of this kind. Alexander Gibson, who had been responsible for the restaurant, designed a series of private houses and house-conversions and acted as architectural consultant to Gilbey's for their distillery and warehouse at Harlow. Kenneth Bayes designed extensions to a school for the Society of Friends at Saffron Walden and, with Misha Black, a new canteen for Sunbeam-Wolseley in Eire. There was also one commission, in 1958, which was of special interest because it had a widespread influence on towns and cities all over the country. This was the revitalization of Magdalen Street in Norwich, a scheme which was neither architecture nor interior design, but drew on an understanding of both.

Magdalen Street was typical of almost any shopping street in an English country town – an untidy conglomeration of buildings, shop fronts, signposts, and street furniture, all at various stages of redecoration or decay. The Civic Trust wanted to experiment with the translation into everyday terms of some of the lessons which the Festival had taught about designing for an urban environment. They chose Misha Black to co-ordinate the project and, with the help of the Norwich authorities, persuaded the owners or occupiers of over eighty properties to co-operate in a scheme which involved a detailed survey of every building in the street. A range of standard colours for walls and woodwork, of materials for curtains and awnings, and of alphabets for signs and fascias, were all set down in a manual for use by the designers, who included local architects; and, at the same time, new designs were made to replace the clutter of street-lighting fixtures, hanging signs, 'no-waiting' signs, 'bus stops and litter bins. Like the Festival, it was a major exercise in organization and, as *Design* magazine commented at the time, it provided 'tangible evidence of an awakening of public conscience against the mediocrity and muddle that has for so long been accepted as inevitable in our streets'.

The project led to a further commission for the rehabilitation of the centre of Burslem and inspired several hundred similar schemes by local authorities in other towns and cities, under the guidance of the Civic Trust.

In 1959, to encourage the expansion of architectural work, DRU

formed a separate office called Black, Bayes, Gibson & Partners; the
'partners' included Anthony Wilkinson and William Apps who have
extended its scope to include not only the factories, houses, and office
buildings undertaken before, but also churches and schools.
Architectural work carried out by Black, Bayes & Gibson has usually
been for projects which have been thought to be better executed
outside the DRU 'umbrella', with its basically design-office image.
More recently, the Unit's architectural activities have been further
extended by a young architect, Richard Rogers, and his sociologist
wife, Su, who have formed an independent practice in association
with DRU.

Possibly the best-known job to come out of these new architectural
off-shoots was the small mammal house at the London Zoo, designed
by Misha Black and Kenneth Bayes in 1961. Bayes has said that his
study of the problem of creating a satisfactory environment for
mentally subnormal people helped in designing the mammal house,
in that both were for 'clients' unable to express their needs, and whose
experiences the architect was unable to share. Research had to be
carried out to find the best possible environment for each animal and,
especially, to ensure that there would be no adverse effects, on their
health or breeding habits, of such things as the reversal of night and
day, which was achieved in a special section for noctural animals,
called 'The Moonlight World'. Here, special lighting gives a moonlit
effect to the cages during the day, so that visitors can see the night
animals while they are active and awake.

Perhaps the most important contribution to the development of design
since the war, however, has been the evolution of services to meet a
growing business need for comprehensive design policies. Such work
has been increasingly significant for DRU, for its activities have
extended beyond the more conventional approach to corporate
identity, to an increasing concern with design as an integral part of
business management.

Before discussing this development, however, it is necessary to
consider other fundamental changes in the philosophy of design for
industry which have taken place during the years since the Festival.

**Design mystique and the
human sciences**

Throughout the period of growing affluence since 1951, products
gradually became more plentiful, and competition for the consumer's
money slowly increased. The more forward-looking firms realized
that industrial design would become progressively more important in
the buyer's market that was emerging. Encouraged by this, and by its
success in selecting products for the Festival, the Council of Industrial
Design began drawing up plans in 1953 for a permanent exhibition
of well-designed products in the centre of London. A proposal of this
kind had been considered by the Gorrell Committee in 1934, but the
Council felt that sufficient progress in the design of consumer goods
had been made to justify a permanent exhibition of the kind which,
before the war, had been rejected because of the difficulty of stocking
and maintaining it.

The Board of Trade supported the CoID's proposal, and the Centre
opened in May 1956, paid for partly by the Government and partly by
the charges made to manufacturers whose products were selected for
exhibition. The Centre attracted immediate interest from the public, and
attendances were soon averaging 2500 a day, later to be increased by a
further 1000. For the first time the Council had a powerful channel of
communications direct to the consumer. Apart from occasional
exhibitions, its work over the previous twelve years had been directed
to the means of producing and distributing products of improved
design. Priority had been given to a need for the better education of
designers, their more widespread employment in industry, and to a
greater awareness and acceptance by retail buyers of the new designs
which industry was introducing. Even *Design* magazine, which had
been started in 1949, was aimed primarily at industry, so that
information for the public, before the opening of the Centre,
could generally be dealt with only by feeding material to the consumer
Press.

Although the Centre brought widespread recognition and acclaim
for the CoID, all was by no means sweetness and light. Many
manufacturers could not see that, by selecting products of an advanced
style, the Council's aim was to influence consumer choice, and thus to
create a market demand for something better than the normal run-of-
the-mill merchandise. That some of the products selected in this way

were ahead of mass market acceptance, and therefore appealed to a sophisticated minority, led to the idea that 'good design' does not sell. And when products were picked out for special commendation in the series of annual design awards, which started in 1957, the notion soon got around that to win an award was the 'the kiss of death' commercially – at any rate in the opinion of those firms which did not get one. Fortunately the award-winning products which fitted this particular brand of pessimism were comparatively few in number, which suggested that the public was more forward-looking than many firms realized.

A more serious problem for the Council arose from the growth of the consumer protection organizations which established, through *Which?* and *Shopper's Guide* magazines, that the performance, quality, and reliability of products could be compared on a scientific basis in controlled laboratory tests. This factual, down-to-earth approach appealed to a wide section of the intelligent public, but it ignored the aesthetic aspects of product evaluation since these were not so susceptible to measurement. In contrast, the achievement of a satisfying visual effect had always been high among the priorities of the industrial designer, and was an important consideration in the selections made by the CoID. This more intuitive approach to product evaluation led to the criticism that products accepted for Design Index could well be functionally less effective than those which were turned down. The public, it was argued, was being misled and, in a scathing attack in *The Spectator*, Reyner Banham described The Design Centre as 'H.M. Fashion House'.

The Council responded to this challenge by introducing a series of checking procedures to ensure that accepted products satisfied minimum technical standards, but this did not entirely overcome the feeling of mistrust of industrial design, particularly among some members in the engineering industries.

Because of this mistrust, and because Britain is always assumed to be a nation of Philistines as far as aesthetics are concerned, those involved in promoting industrial design tended to base their arguments more on the achievement of such practical results as ease of use or economy

of production than on improvements in visual quality. There was
even a tendency to downrate the importance of aesthetics, or at
least to leave it out of the discussion, in order to make the idea
of industrial design acceptable at all to the 'hard headed' engineer or
business man.

The effect of this attitude over the years was for the aesthetic
aspects of industrial design to develop into a mystique. It was almost
as though 'good design', at least as far as aesthetics were concerned,
was a special kind of look or style which you either liked or you didn't
– and if you didn't you were on the outside, and could be spoken to
only in terms of simple, practical things appropriate to the uninitiated.
Thus, during the 'fifties, the idea had become established that the
visual aspects of a product were matters which should be left entirely
to the designer whose own taste and judgement were 'correct', even
though the consumer might have quite different views.

A growing awareness of the human sciences, however, provided at
this time the foundations upon which fundamental changes in the
philosophy of design for industry have subsequently been built. It
gave rise to a new emphasis on the use of logical principles in design,
and enabled the problem of aesthetics to be seen as part of a wider
concern with human values in the total design mix. As such it has
taken a good deal of the mystique out of industrial design and has
helped to increase an understanding among industry and the public of
the special functions which the industrial designer performs.

Ergonomics had the most direct influence on design thinking, since it
encompassed a number of specialised aspects of human science
including anthropometry, physiology, anatomy, and psychology, all
of which could be seen to be directly relevant to product design.
Ergonomics has influenced design in three main directions. First, it
emphasized that, in the design of products, the needs of consumers
were more important than those of the designer; second, it put
forward the, then, rather novel idea that such needs can be determined
by scientific means, rather than by guesswork or intuition; and third,
it provided a systematic framework for the design activity.

The importance of ergonomics, of fitting machines to the capabilities of the people operating them, was first recognized during the war, when physiologists used their knowledge of the structure and functions of the human body to help engineers to lay out the cockpits of aircraft. The pilot of a Hurricane or a Spitfire could not afford to fumble for a switch or sit in discomfort. Everything had to be immediately and easily accessible because the consequences of error might well be fatal. Later a Swedish scientist, Bengt Akerblöm, demonstrated how information about human body sizes and anatomy could be applied to the design of seating, leading to notable departures from traditional chair forms. Other scientists working in similar fields built up on this early work, and they came together after the war to form the Ergonomics Research Society which did much to spread interest in the subject among designers and engineers.

At first many industrial designers were reluctant to accept that ergonomics had anything to add to what they were already doing. In many cases solutions to problems devised by an ergonomist were described as common sense. What is interesting, however, is that while the result was seen to be common sense once the solution had been provided, the common sense was by no means so obvious at the start.

To the industrial designer who had successfully pioneered *avant-garde* attitudes to design in industry only a few years before, the idea of a scientist coming along with answers based on statistics seemed presumptous. And because human requirements established by the ergonomist sometimes resulted in the need for changes in appearance, it was even assumed that ergonomics was impinging on the industrial designer's own cherished field of visual aesthetics.

This misunderstanding gave rise to endless discussion and argument. A degree of opposition to ergonomics was probably created by those who exaggerated its claims in the early days. Nevertheless it could be shown, without much difficulty, that in countless minor details many products coming onto the market were deficient in the attention given to the way in which they would be used. In the consumer goods field it was found, for example, that tables and working tops were too high, that beds were too short, that knobs were difficult to reach, dials impossible to see from certain angles, and lighting fittings were put in the wrong places, spreading shadows where light was needed.

But while these faults in design were irritating to the consumer, they were seldom disastrous. Studies of car design, on the other hand, suggested that the consequences of poor ergonomics were a good deal more serious. One driver described how, at 40 mph and in a stream of traffic, he pushed a button expecting to work the wind-screen washers, but instead pushed an identical one beside it which opened the bonnet.

Analysis of fascia controls on some cars showed them to be out of sequence and occasionally impossible to reach from the driving seat, especially if a safety harness was being worn. Seating adjustments, head-room, the arrangements of foot controls, and other aspects of interior layout were found to be unsuitable for a great many people, particularly the shorter and taller sections of the adult population. Studies of production machinery, power station control rooms, and many other situations where people spend long working hours at specific tasks revealed a similar lack of attention to the physical and psychological needs of the worker.

The cult of the 'average man', often embodied in a two-dimensional cardboard manikin, resulted in more bad design than almost any other cause. It was particularly harmful because it was used, mistakenly, in the belief that it represented an ergonomic approach. It has taken nearly twenty years to lay the ghost of the average man, and even now this phantom still returns to haunt us from time to time.

But while the mistakes were often simple enough to spot, the solutions were not always so easy. New research has shown how little we really know about human behaviour. Designs based on physiological data alone were sometimes found to produce unexpected psychological and emotional reactions which are far more difficult to understand. Nevertheless, increasing attention was given to ways of using psychology to provide more accurate assessments of people's subjective reactions to products. Motivational research is possibly the best known example, while another is the use of perception psychology as a means of solving problems in the field of visual communications. Techniques were developed to study the effectiveness of package designs in the competitive context of a supermarket shelf, and to improve the readability of print, particularly for things like technical publications where information is sometimes required quickly.

More recently, ergonomics has been used in the design of complex industrial control systems, where the inter-action is being studied of one component on another, and the effect of the total system on the human being. We will return to this point later in the book.

Design and management

Seen in retrospect, the direct influence of ergonomics on the design of products, especially in the consumer goods field, has been far less than its protagonists would claim for it. More important has been the influence it has had on the whole way of thinking about designing for industry which emerged during the 'sixties.

In its simplest terms, the change has been an acceptance that a scientific approach in the widest sense is important not only in the provision of information on which design depends, but also in the design process itself. The basic lesson which the ergonomists taught was that the process of finding the right solution to a human factors problem could be ordered in a step by step sequence which was designed to ensure that nothing was overlooked. The lesson was understood by a few research workers who were interested in developing a similar approach, first to the wider issues of industrial design, and later to a comprehensive theory of design embracing all technical, aesthetic, and ergonomic factors.

Articles on systematic design methods began to appear in *Design* magazine in 1958. The foundations were laid by several people who, in the early stages, were working independently but along broadly similar lines. L. Bruce Archer, John Christopher Jones, and John Page in Britain, and Christopher Alexander in America, with a number of others working in more specialized fields, were all developing their own systematic design theories which in many cases had grown out of the procedures developed earlier by the ergonomists.

Archer's work in this field began while he was teaching at the Central School in the 'fifties. Later, he was able to develop his ideas at the Hochschule für Gestaltung at Ulm, and, on his return to England a year later, was appointed by Misha Black, who by then was Professor of Industrial Design (Engineering) at the Royal College of Art, to head a Research Unit where he was able to test his theoretical approach in a series of realistic projects, including a general purpose hospital bedstead. His receipt of a Kaufmann International Design Award in 1964 enabled him to develop his ideas in a book which was to become the first doctoral thesis for the Royal College of Art in its new status as a university institution following the report of the Robbins Committee in 1963.

Jones's interest in systematic design started on a more practical basis, again in the early 'fifties, when he began to carry out experimental studies in ergonomics in connexion with his work at the industrial design office of Metropolitan Vickers in Manchester. His work developed on less rigid lines than that of Bruce Archer, and in 1963 he was able to give more time to this theoretical approach to the subject when he was appointed to start a new department of industrial design technology at Manchester University. He was primarily interested in exploring the possibility of a multi-disciplinary approach to the solution of design problems, and sought postgraduate students from a wide variety of backgrounds.

A new concern with systematic methods in the design of buildings stemmed largely from the appointment, in 1960, of John Page as Professor of Building Science at the University of Sheffield, where he drew attention to the apparent preoccupation in architecture with the appearance of functionalism rather than with buildings which really worked well for their inhabitants.

Alexander's work was concerned more with the application of systematic methods to the design of cities, and he, too, won a Kaufmann award for his writing on the subject.

They all came together at Imperial College in 1962 for the first conference ever on systematic design methods. This conference was a landmark in the history of design, for it pointed the way to the idea of design as an activity in which all the specialists involved could work together on a basis of common understanding. Writing about the conference at the time, *Design* magazine said: 'One outcome of this attitude is the attempt to break down the barriers which more and more seem to divide society – scientist from artist, artist from designer, designer from engineer, engineer from the common man – and so on in increasingly vicious circles. Design is seen as the vital, missing link between these sectors; and the service of man is acknowledged as the primary, overriding objective against which the results must always be measured.'

The death knell of 'art and industry' had begun to toll. What took its place was not so very different in intention from what the Egg exhibit at the *Britain Can Make It* exhibition had set out to explain. The development of systematic design methods, however, took a good deal further the analytical approach which the Egg display had demonstrated. Its aim was to provide a precise definition of design objectives and a means of achieving them which reduced guesswork to a minimum. It adapted a host of techniques from other disciplines to help sort out priorities, reconcile conflicts and map out a course of action which would lead most directly to the desired goal. As part of this process it was concerned to separate the characteristics of a product which could be measured from those which could not, so that the designer should not waste time trying to find intuitively answers to problems which could more readily be obtained with a slide-rule.

Many designers were as sceptical of these new ideas as they had been when ergonomics first appeared on the scene. They criticized some of the procedures which were published as being unnecessarily complicated, and apparently more concerned with their own perfection than with their usefulness as a working tool in the real-world situation in which designers must operate. But the growth of systematic design methods was not an isolated phenomenon. It was directly in line with a parallel revolution that was taking place in management. Management sciences, which had first been studied in America, were to become increasingly important in Britain throughout the 'sixties. Techniques like network analysis, operational research or discounted cash flow were being introduced to cut out the uncertainties in business operations so that company objectives could be accurately identified and achieved. The similarity with new ideas about design was not fortuitous. That people working in fields apparently so widely separated as management and design should be working on the same lines pointed to a broad climate of ideas in which design could be seen as an integral part of industrial management.

A decade had elapsed since the first tentative suggestions were made in the CoID's 1951 design congress that design is a responsibility of top-level management. By the mid 'sixties the way was clear to a firmer understanding between the two. The techniques and terminology of design had been brought into direct alignment with those of management. Both had been influenced by the development of computers, not only in the sense that the computer could provide information on a scale which previously would have been impossible, but also because it provided a language and a logical structure of thought which scientists, engineers, managers, and designers were increasingly to share.

With a common framework, the integration of the design activity with the general management problem became much clearer than it had been before. It could be demonstrated that, to be effective, the design process was dependent on decisions across the spectrum of management activity, spanning marketing, production, sales, and overall financial policy. Yet to siphon off relevant information from these sources, and to interpret it in a form that would be useful to the designer, required a special expertise which hardly existed at all in industry. The missing technician of the 'forties had been the industrial designer. In the 'sixties the missing technician was, and to a large extent still is, the manager of the design process.

Many industrial designers during the intervening years had found that the biggest problem they faced was the extraction from their clients of a comprehensive and realistic brief. Often designers found they had to prepare their own briefs because there was nobody within the company with the necessary skill to do it for them. The idea of design management thus grew from this understanding that the design process itself was failing to connect satisfactorily at all points with the wider activity of company management.

Corporate identity and total design

Within the context of this new interest in design and management, the development by DRU of its graphic design services has a special significance. Even in the comparatively early days, Milner Gray saw the need for a more systematic approach to the problem of obtaining from his client all the information necessary to enable a comprehensive brief to be written. Over the years he built up a check list of some 150 questions, and has been known to work systematically through these in order to establish his client's precise requirements.

The use of such a check list, however, had other virtues, for it meant that companies were forced to think about problems which they had not previously considered. What might have started as a simple commission to redesign a firm's stationery could be shown to be only part of the problem, and sometimes by no means the most important part. The idea that design could be used as a business tool, to express a firm's unique identity through all aspects of its operation, had existed in theory before the war. But its realization was very much a post-war phenomenon, and had developed on a large scale only during the past ten to fifteen years.

DRU's first involvement in a comprehensive design programme of this kind was in 1946 for Ilford, the manufacturers of photographic materials and accessories. It established a relationship which has continued ever since, and was the beginning of several long-term consultancies for which Milner Gray has been responsible. In this way, the Unit's graphic designers have provided advice on the visual implications of such things as trade-marks, stationery, packaging, catalogues, signposting, vehicle liveries, and uniforms as part of a joint team with the Unit's architects, interior and product designers who have been brought in as necessary to deal with the design of shops, showrooms, offices, factory layouts, and products. Projects of this kind were carried out for Courage and for Watneys, the brewery companies; Gilbey's, the wine merchants; the Hodder Group of publishing companies; Dunlops; Capper Neil, the building contractors; the multi-product company of Thomas de la Rue, and many others.

In the early post-war years such schemes were referred to as 'house-styles'. The description was appropriate, for it referred to the establishment of a recognizable visual image for an individual firm. The structure of industry, however, began to change rapidly during the 'sixties. Companies swallowed each other like fish in a river, or they linked themselves together in ever more complex patterns of organization. In such conditions the term 'house-style' no longer fitted the new problem of creating a common identity out of a conglomerate association of different firms, each with its own character and methods of work. What was needed was a 'corporate identity', and this term, borrowed from America where the restructuring of industry had begun some years earlier, was now widely adopted in Britain.

From the designer's point of view, the difference between 'house-styles' and 'corporate identity programmes' is one of scale. This is reflected less in the amount of design work which has to be done than in the problems of management and organization. Few large companies appreciate at first the full implications of a decision to go ahead with a corporate identity programme. To most of them, it is considered as a cosmetic operation, with the creation of a rather more attractive image by adroit use of tweezers, eye-shadow, and 'cake' foundation. What they overlook, as any beauty expert could have told them, is that no amount of surface make-up can rejuvenate an ageing face. Corporate identity programmes have thus helped to focus attention on more fundamental aspects of business health, and this is directly in line with the wider concern in industry with the efficiency of its management. There is little point in having a letter-heading that evokes an impression of an efficient, forward-looking company if the letter written on it is badly typed, rude, or illiterate. Nor is there any value in a luxurious-looking pack for a poor quality product.

While these may appear to be comparatively minor issues in themselves, they emphasize the need for a corporate image to reflect the actual achievement of a corporate attitude throughout the management hierarchy of a company. They also affect more important questions which are central to management policy. A group of companies could well find that corporate identity is in conflict with its policy of competition among its subsidiaries. In these circumstances corporate identity may not be required at all, and design programmes of a different kind would be needed. Whatever the circumstances may be, it is essential that a company's management objectives are clearly defined at the start of design activities, for the objectives determine whether the design programme will be concerned, for example, with changes in marketing tactics; improvement of public relations; cost reduction through rationalization of stationery and equipment; the up-grading of product policies; or a combination of several or all of these. In some cases, design work may extend far beyond the graphic orientation of most corporate identity programmes into the design of interiors, products, buildings and many other aspects of a company's activities. The term 'corporate design' aptly expresses the nature of such extensive design programmes, and several examples by DRU are illustrated in this book.

The difficulties of extracting from a large corporation a precise outline of its design requirements are very much more complex than those for a simple house-style for a single firm. A straightforward check list will not alone be enough, and other techniques must be devised.

Design Research Unit has evolved its own pattern of investigation. Faced with the problems of carrying out corporate design programmes for some of the largest industrial organisations in Britain, both Milner Gray and Misha Black believe that it is essential that initial discussions should take place at a top management level, so that the major policy issues can be resolved at the start. From such discussions a series of questions for management can be formulated which the company may be asked to think about so that the main lines of the brief can be set out in broad principle. Since the subsidiaries and associated companies of large organizations often have a considerable degree of autonomy, there is no military chain of command enabling orders issued at the top to be carried out unquestioningly by those below.

It is therefore necessary to set up a committee or working party in which representatives of the associated companies can participate in a mutually agreed plan. These committees have two main functions: the first, and perhaps the most important, is to examine and if necessary reinforce the need for corporate action; and the second, to provide a communications channel through which information can be collected and reactions to initial design solutions tested.

Having set up the administrative machinery, the next stage is an on-the-spot investigation by members of the design team, sometimes accompanied by the client's own representatives. Such fact-finding surveys can be extensive. For the corporate identity programme which DRU was commissioned to design in 1961 for the British Transport Commission – as it was then called – Milner Gray and Kenneth Lamble, with the Commission's assistant design officer, visited seven European countries in connexion with the redesign of staff uniforms alone. A more recent project for Bata International involved, as well as the usual graphic design problems, the redesign of the company's shoe-shops all over Europe. In the investigations for this, two of DRU's graphic designers, Ronald Armstrong and June Fraser, and two of its architects, Misha Black and Kenneth Bayes, visited 114 shops in eight countries.

The accumulated information from these investigations, and from discussions with the client, is analysed and the results set out in a report which defines the problem and proposes appropriate solutions and working procedures. These reports, which for a job of major complexity may well be in the region of 30,000 words long, are intended more to record an agreed stage in the progress of the work than to present the client with totally unexpected proposals. They therefore provide the basis for a detailed understanding between client and designer from which design work itself can proceed through all its stages, including provision for implementation and maintenance.

In 1968, DRU was commissioned under Milner Gray's direction to prepare a corporate identity scheme for ICI, Britain's biggest commercial enterprise. At the time of writing, design work for this project was at an early stage, and therefore cannot be fully discussed or illustrated here. But the work so far, among ICI's eleven companies, is an outstanding example of the careful planning which goes into large-scale programmes of this kind.

While DRU would not claim that the procedures it has evolved are unique in themselves (many design offices today provide house-style and corporate identity services), it is the scale of the operations with which the Unit has largely been concerned that has made it necessary for DRU to develop its special expertise in the management of these programmes, and the existence of this expertise has enabled the Unit to pioneer a new kind of design service which may well become increasingly important in the future. This is essentially a form of management consultancy, but it is concerned here specifically with detailed investigations of a company's design activities leading to recommendations on how its design policy can best be organized.

The first real opportunity for this type of independent investigation came when, in 1964, the Co-operative Wholesale Society asked DRU 'to survey the present methods by which decisions concerning the design of products and packaging are made by the CWS, and the facilities available for their implementation; to advise whether the present procedures result in the production of designs which best serve CWS commercial interests; and to make proposals for methods of improving the design standards of CWS products in relation to its marketing needs.'

To carry out the investigation, a working party of seven DRU members, led by Misha Black, arranged a programme of visits with the object of providing comprehensive design information through all the ramifications of the CWS organization. The team visited twelve factories, eighteen retail shops, and three salerooms, besides holding discussions with people concerned with advertising, architecture, market research, and technical research, as well as managers and buyers all over the country. What resulted from this mammoth survey was a picture of an organization that had grown far beyond its capability to plan and implement an effectively co-ordinated programme of design work, and the main recommendation was for a centrally directed design policy fully supported at board level. Among the detailed recommendations was the appointment of a design panel which would advise the board on all matters affecting the design of cws products, packaging, print, architecture, and all other aspects of graphic and industrial design. In the event, these proposals for a fundamental reappraisal of its design activities were only partially accepted. The appointment, however, of a design officer at the beginning of 1966 fulfilled one of the recommendations, and it is to be hoped that in time it will lead to the acceptance and implementation of the others.

Whether design management consultancy of this kind is best carried out by a firm which itself provides design services is open to question. The danger is that, because of its preoccupation with design, it will not be in a position to look at a company's design activity objectively within the total context of a company's management problem. Ideally, advice on the management of a company's design policy should be a part of the services provided by management consultants. Unfortunately, however, the management consultants, and the business schools with which some of them are closely associated, have paid little attention to the need for proper organization of the design process in industry. Consequently they lack both the interest and the knowledge to extend the services they already offer into this new and important field. Until this situation can be changed, the designers themselves are the only people qualified to develop and advise on this specialized branch of management.

Management implications for DRU

Understandably enough, this concern with the management of the design process has led DRU to examine its own efficiency as a business organization. In 1967, a firm of management consultants, Urwick Orr and Partners, was brought in to advise on the Unit's problems. Over the years DRU had grown to meet the demands of the more comprehensive and complicated commissions of the kind we have referred to above. The increase in the organizational aspects of the service provided by the Unit had led to an uneconomic relationship between its design and administrative staffs, and this, among other things, had resulted in profit margins that were too low to ensure a stable and healthy continuity. Of the sixty members of staff which

DRU employed by 1967, a third were concerned with clerical and administrative activities. At the same time the division of the Unit into its architectural and graphic design sections had led to separate costing procedures and a certain lack of overall direction, which the management consultants thought had made it difficult for DRU to use its total resources to the best advantage. Over and above all this, however, was an awareness that the senior partners were reaching an age when at least partial retirement would soon have to be faced. In these circumstances a more formal organizational structure for the Unit was needed to provide a basis for planned continuity and growth in the future.

Urwick Orr reported that it was necessary to the professional character of the Unit that all partners should contribute to the policy governing its affairs, and should each have some responsibility for profitability. They recommended a reconstruction of the Unit to make this possible – three groups to replace the two which had existed before. At the head of each group would be one of the younger partners who would have a specific responsibility for the efficient running and financial success of his group: Anthony Wilkinson for interior design and architecture; Christopher Timings for graphics; and James Williams for a newly created product design group. Above this, the ultimate direction and control of the Unit would be the responsibility of a reconstructed board. Procedures were suggested to ensure full consultation between the board and the partners in the planning of future policy, and the co-ordination of these managerial functions was to be reinforced by new centralized systems of accounting and cost control.

Urwick Orr's recommendations were accepted and put in hand straight away. DRU had had too much experience in its own field of consultancy to doubt the value of outside expert advice in bringing a fresh eye and a wider experience to bear on problems which people within an organization are often too closely involved with to see in clear perspective.

Now that firms are becoming more knowledgeable about the function of design in industry, it is making many of them realize that there are

64

advantages in 'shopping around' for those design services which most efficiently supply their needs. And, as we shall see in the final chapter, the challenge in the future to design groups like D R U will be the demand for an even greater emphasis on skills in planning and organization.

Interior design

All the illustrations on the following 48 pages show examples of work carried out by Design Research Unit during the past 25 years. Except in special cases, individual credits to the designers responsible have not been included, but credits to architects and others with whom the DRU designers have collaborated are given, where appropriate. The illustrations are divided into five sections and the first of these, on interior design, shows a field of design activity in which DRU has been prolific and successful, particularly during the last decade.

1952
One of the first major interior design projects after the *Festival of Britain* was for BOAC's new headquarters at London Airport, Heathrow. The illustration shows one of the work benches in the instrument and radio shop. The benches were designed after intensive studies of ergonomic and functional requirements, and prototypes were constructed and tested before the designs were put into production. The architects for the building were Sir Owen Williams and Partners.

1959
Design of the public areas for the P&O-Orient Lines' ship *Oriana* included the first class lounge, *below left*; and a children's playroom, *below*, incorporating a climbing frame in teak with aluminium supports. A special version of the ship's badge is shown on the bows, *left*.

1952
Main staircase of the Time & Life Building, *opposite*, the interiors of which were a pioneering exercise in commercial luxury. Designed in association with Sir Hugh Casson. Architect: Michael Rosenauer.

1962
Opposite: Entrance hall of the headquarters building for *The Times* in Printing House Square, London, part of a project involving the design of major circulation areas, an executive suite, a book shop and general consultation on lighting, finishes and sign posting. The architects were Ellis Clarke & Gallannaugh with Llewelyn-Davies, Weeks, Forestier-Walker & Bor as consulting architects.

1962
The three areas, *right*, are from a comprehensive interior design scheme for the Co-operative Insurance Society headquarters in Manchester. This is a 400 ft high office block, designed by the architects Sir John Burnet, Tait & Partners. *Top*, the west vestibule and entrance hall on the twenty-third floor. *Centre*, the control room for heating, humidity and other services with specially designed services diagrams, and desk control panels. *Bottom*, the boardroom on the twenty-third floor.

1962
Two examples from an experimental series of
modernized branch post-offices in London
designed in association with Sir Hugh Casson.
The illustrations, *left*, show the Knightbridge
post-office before and after redesign. The
modernized branch office at South Molton
Street is shown *opposite*. The aim of the
project was to develop a range of standardized
exterior components to provide an instantly
recognizable style.
Interior equipment such as counters, signs,
form dispensers and telephone kiosks were
also re-designed.

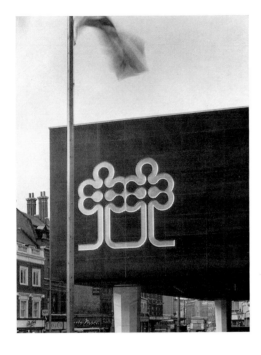

1965
The foyer of the Royal Garden Hotel, Kensington, *opposite*, was part of a project involving the design of the hotel's public areas. The symbol, *above*, creates a focal point on the exterior of the building, which was designed by the architects R. Seifert & Associates.

1964
Right: Two views of interiors at West London Air Terminal. *Right, above,* the large reservations hall, incorporating specially designed desks for the automatic seat-reservation system, carried out in collaboration with BEA. *Right, below,* is a view of the passenger check-in desks. The architects for the building were Sir John Burnet, Tait & Partners.

1967
All the illustrations on these two pages show interiors of Britannic House, the London headquarters of British Petroleum. Management control was carried out by a design panel representing BP; Joseph F. Milton Cashmore & Partners, the architects of the building; and DRU, who were responsible for the interiors of the special areas. *Left, top,* is the staff restaurant on the lower ground floor which seats over 400 people, with a further 150 on a mezzanine floor. The mural is by Edward Bawden. *Second from top,* is a committee room on the thirty-second floor with photographs of past members of the board incorporated in the panelling of the end wall. *Third from top,* is a cinema forming part of a conference suite on the ground floor. The illustration, *bottom,* shows a typical director's office.
Opposite is the entrance hall with a steel relief by Robert Adams above the reception desk.

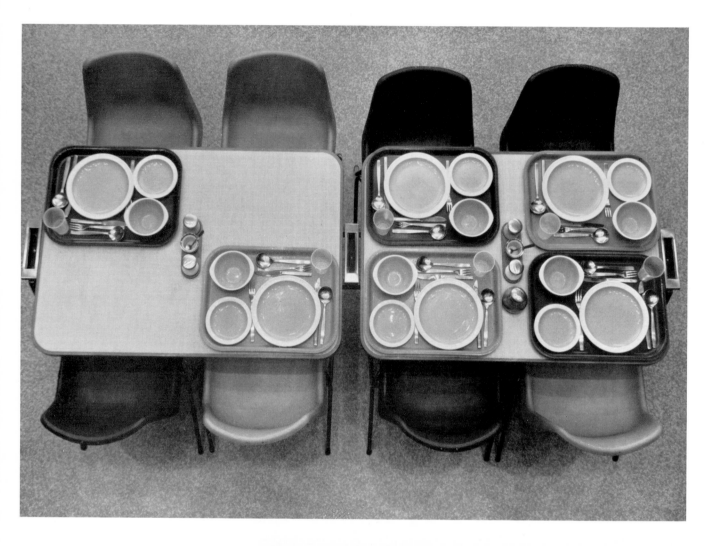

1965
The illustrations, *opposite*, show views of one of the three restaurants at the Staff Luncheon Club for the London headquarters of Barclays Bank in Lombard Street. The table layouts, *top,* have been designed as an integrated system of utensils, trays, and table top, with fixed ash trays and condiment holders acting as location guides for the trays.

1968
Two views, *right, top and centre*, of the *Ohrmazd*, a cargo and passenger ship owned by the East & West Steamship Co., Karachi, for which DRU were consultant designers for the superstructure and interiors.
One of the passengers decks is shown, *top*, and the passenger lounge, *centre*.

1968
Right, bottom: The Council Chamber in a reconstructed part of Marylebone Town Hall, now the Westminster Council House, London. The councillors' desks incorporate a push-button voting system. The architects for the reconstruciton were T. P. Bennett & Son.

Exhibitions and architecture

Like other design offices at the time, DRU's early post-war work was largely concentrated on exhibition design. The climax of this activity came with the *Festival of Britain*, and there was a marked decline during the decade which followed. While architecture as such has never formed a major part of the Unit's work overall, it has always been workmanlike and occasionally distinguished. It has also been significant in its contributions to interior design, townscape and specialized jobs such as the small mammal house at the London Zoo.

1946
The giant egg symbolized the theme of a special exhibit at the *Britain Can Make It* exhibition, designed to illustrate the role of the industrial designer. Even an egg-cup, it was demonstrated, creates marketing, production and other problems which the designer is trained to solve.

1968
Opposite: Entrance hall for the Chase Manhattan Bank in Mount Street, London. The design generally follows the bank's house-style, established by the American architects Skidmore, Owings and Merrill.

1948
Design at Work, an exhibition at the Royal
Academy of work by members of the faculty
of Royal Designers for Industry, organized
by the Council of Industrial Design and the
Royal Society of Arts. The central feature of
the transport section was an experimental
jet aircraft, the D.H.108.

1951
The *Festival of Britain* on the south bank of
the Thames. The site was divided by the
Hungerford railway bridge, and the section
at the top of the picture was co-ordinated by
Sir Hugh Casson, James Gardner and
R. T. James. The section at the bottom of the
picture was co-ordinated by Misha Black,
Ralph Tubbs, James Holland and Ralph
Freeman. The drawing, *bottom left*, shows an
early proposal by Misha Black for a pavilion
to house an international exhibition on the
south bank.

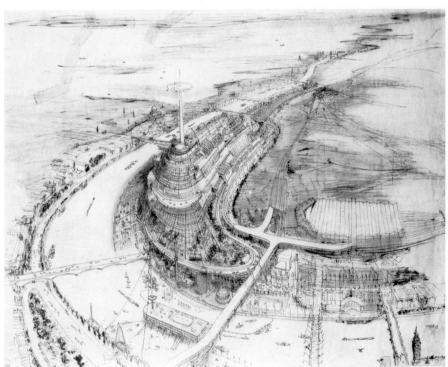

1951
Opposite: Four views of the *Festival of Britain.*
Top, left, the upstream area at night, with the
Skylon, designed by Powell & Moya.
Centre, The Regatta Restaurant, *left*, and part
of the sign-posting scheme, *right. Bottom*, the
interior of the Dome of Discovery. Ralph
Tubbs was the architect for the Dome.

1957–61
During these five years, DRU acted as
co-ordinating designers, in association with
John Bruckland, for the Furniture Exhibitions
at Earls Court, London. The illustration,
right, top, shows the central feature of the
1958 exhibition which consisted of a curtain
of plywood panels. *Right, bottom*, is a special
feature for the 1960 exhibition, a display of
paintings, drawings, prints and sculpture
called *The Artist in Your Home.*

1959
Top: This extension to the Friends School at Saffron Walden, in Essex, forms a courtyard with the old buildings which can be seen on the right of the picture.

1955
Centre: The house in Cannon Lane, London, has a first floor terrace to make up for the small size of the garden. Materials and detailing were chosen to harmonize with the eighteenth-century house behind it.

1956
Bottom: Part of an Anglo-Jewish art and history exhibition at the Victoria and Albert Museum, London.

1960

These illustrations show the rehabilitation of Burslem's town centre. This was the second project of the kind carried out for the Civic Trust, and was undertaken to mark the fiftieth anniverary of the amalgamation of the Five Towns. The picture, *top*, shows the town centre before the scheme was started and, *centre*, part of the same area after it was completed. The work involved the re-routing of traffic, the creation of a garden area, the planting of trees, extensive re-painting of the surrounding buildings and the introduction of new street furniture and sign posting. Before and after examples of the latter are shown *bottom left and right*.

1967
Entrance hall, *top left*, and south-facing exterior, *bottom left*, of the Alliance Building Society's head office at Hove. The offices were built on an 11½ acre site and were planned to take advantage of the surrounding landscape. Open planning was used for the office areas and flexible internal arrangements were designed to allow for later expansion. DRU were the consultant architects and interior designers. The architects for the building were Jackson, Greenen & Down.

1967
Child's eye view of a civet, *opposite*, seen through the plate glass window of a cage in the Charles Clore Pavilion for Mammals at the London Zoo. This was the first house of its kind especially designed for small animals, and extensive research on environmental requirements were carried out in collaboration with the Zoo's staff. Further illustrations of the pavilion are shown on the next two pages.

1967
Further views of the small mammal house.
Opposite, top, dark walkways were designed to
focus attention on the natural day-lighting of
the cages. Splayed glass minimizes reflection
and dirtying. *Centre:* A group of cages,
far left, in which non-jumping animals are
separated from the public only by a five foot
wide water barrier and a warm air curtain
and, *left,* a service corridor behind the cages
showing water bottles, nesting boxes and
access doors or grilles. *Bottom, left:* Entrance
to the pavilion from the Regent's Canal Bank.
Right, top: A 'conservatory walk' hung with
trailing plants in suspended troughs. *Right,
bottom:* Entrance to the basement section, *The
Moonlight World,* in which special lighting was
installed which reverses day and night for a
group of nocturnal animals, so that visitors
can see them while they are active and awake.

Graphic design and corporate identity

Major changes have taken place during the past two decades in the provision of graphic design services for industry. The demand for limited display, packaging or other point-of-sale projects has gradually given way to wide-ranging house style and corporate identity programmes designed as an integral part of a company's marketing strategy. The trend has followed the restructuring of industry into larger and more complex units, and has led to the need for increasing organizational skill by the design profession, and an acceptance of more systematic managerial disciplines.

1951
Decorative glass screen at the entrance to the royal box at the Festival Hall, London. The rendering of the coat of arms was carried out on the reverse side of the $\frac{5}{16}$ inch glass by deep engraving and acid etching to create an appearance of strong three-dimensional modelling.

1950
Three illustrations showing a house style designed to improve the visual identity of Courage & Co. Ltd and to add to the amenities of its inns and public houses. *Above* is the re-drawn symbol designed to be adapted for use in a variety of materials and in a wide range of sizes. The symbol can be seen, *right*, on a range of specially designed stoneware jugs; and, *bottom right*, on a range of beer bottle labels.

1952
Left and below left: House style for Tate & Lyle shown here in its application to the transport livery and packaging.

1951
Below: Bottle and label for Eno's fruit salts, before and after re-design.

1954
Opposite: Symbol for Austin Reed, shown here engraved on the marble fascia of the Regent Street store, London. The symbol was part of a wider design programme for the company, including a house style covering print, packaging and the company's transport.

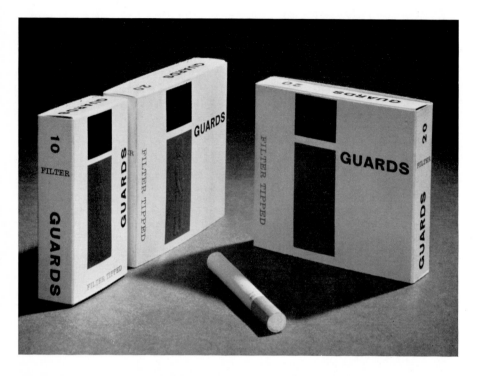

1958
The pack for Guards cigarettes, *opposite*, in red, black and gold, with the figure of the guardsman blind-embossed on the front red panel, has proved to be an outstanding commercial success.

1960
The bottle labels and printed cans, *opposite*, continue the character of the original designs for Guinness labels.

1961
The trade mark, *opposite*, was designed for the Midland Metal Spinning Co. Ltd, and symbolizes the 'Tower Brand' name under which the company's products are marketed (see also page 102).

1962
The illustrations, *right*, show two applications of a house style designed for Dunlop Footwear Ltd. The 'foot-mark' is used as an overall pattern on the packaging and has been designed in white or black on a green ground.

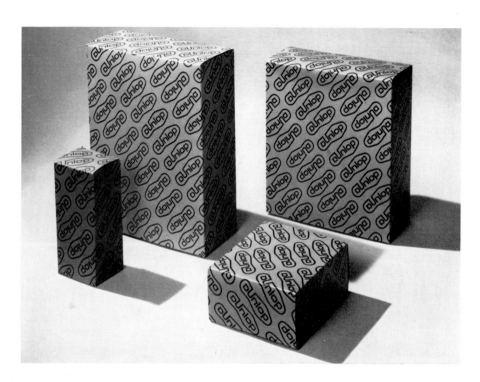

1965
Opposite: The change from earlier house style projects to more complex corporate identities is illustrated by this scheme for Tarmac, the leading company in a group of seven. The scheme was designed for eventual extension to all the companies in the group, a unification which is expressed in the symbolic cluster of seven 'T's'.

1962
Right, top: A range of stocking packs for Berkshire Knitting (Ulster) Ltd.

1964
Right, centre: Two packs from a house style programme designed for Josiah Wedgwood & Sons Ltd, using the traditional Wedgwood decorative motif.

1968
Right, bottom: Unit packs for punched cards for Hoskyns Systems Research. The packs are delivered flat for assembly by Hoskyns' staff who insert the cards and identify each pack by a stick-on patch under a series of pre-printed numerals on the sides of the packs. They are then sent with four manuals in a transit outer carton.

1968
Left, top: Corporate identity for Capper Neil, a company with autonomous subsidiaries, manufacturing plant for the building and construction industry. The scheme, designed to be introduced progressively over a period of two years, is intended to provide a clear visual link between the subsidiaries, which are normally engaged at any one time on some 250 work-sites.

1968
A major corporate identity programme is now being introduced for ICI to bring a closer visual relationship between the company's eleven divisions. The ICI mark, shown on the flag, *left centre*, has been modified in detail to improve its clarity. A pilot study to test proposals for sign posting and hazard signs has been carried out at ICI's Wilton works, *left, bottom*, to prove their value for more widespread applications.

1965
Opposite: Symbol and typographical style for the Confederation of British Industry.

Product design

Comparatively little individual product design work has been carried out by DRU during the quarter century of its existence. This is largely because the Unit's special contribution to the development of post-war design has been in the field of large-scale corporate design projects in which product design is sometimes an integral part, as in the work for Ilford, British Rail and London Transport, which are illustrated in the next section.

1947
Right, top: Gift boxes in brushed aluminium, designed for the Rolex Watch Co., Geneva. These watchcases were intended to be re-used as pencil boxes.

1947
Right, bottom: Plastic egg-cups designed for Holmes (Wragby) Ltd.

1968
Opposite: New street name plates designed for Westminster City Council. Fixings are concealed and the dirt-collecting frame, previously used, has been discarded.

Opposite: Four products designed between 1947 and 1961.

1947
The gas meter was for the Gas Light and Coke Co., and is shown here in the form of a working prototype. Jack Proctor was associated with D R U in the design of the external casing.

1955
Pyrex oven-to-table glass, for Joblings of Sunderland.

1958
The saucepan, *bottom left,* was in vitreous enamel in three colours and three sizes. They were designed for Ernest Stevens Ltd.

1961
The other saucepans, *bottom right,* were part of a range of heavy gauge aluminium holloware designed for Midland Metal Spinning Co. Ltd, the manufacturers of 'Tower Brand' cooking utensils whose symbol is illustrated on page 94.

1960
Right, top: School desk and chair in beech and heat-formed ply, designed for Mann Egerton & Co. Ltd.

1961
Beagle executive aircraft – the first flying prototype. The consultant designers were responsible for advice on the interior of the passenger space, details of external form, and the livery.

1966
Left: The two illustrations at the top of the page show a design programme for South Wales Switchgear Ltd aimed at improving the appearance, installation and maintenance of sub-stations. The picture at the top shows a typical installation before the designers were called in and, below it, one of the new sub-stations resulting from the design study. The new substations have been sold to British electricity boards, and have created wide interest in export markets.

1965
Left, bottom: Hose nozzle, shown before and after re-design. In the new version, ABS plastics has replaced metal, decreasing the weight and improving performance.
The nozzle was designed for Dunford Fire Protection Services Ltd.

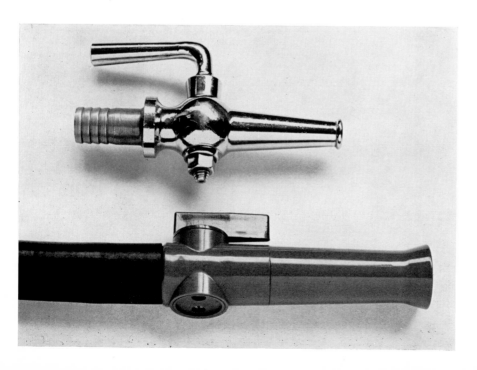

Corporate design

Corporate design is a term used to describe the application of a planned design policy to all aspects of companies' activities, including their products, packaging, graphics, architecture and interior design. Such all-embracing design policies are rare in Britain, for they involve an unusual degree of determination and managerial skill to bring about. The following pages show five examples of corporate design programmes to which DRU has contributed in large measure. Co-ordinated design programmes of this kind are likely to become increasingly important in the future.

1960
Right: Examples from a comprehensive design programme carried out for the Watney Mann Group. The scheme involved a house style, the design of bar equipment and accessories, and an extensive programme for the exterior treatment of public houses, their signs and signposting in conformity with a house identification manual. Some examples of the latter are shown on the next page.

1960
Repainting of Watney Mann public houses.
Right and centre right is the Prince Albert,
London, before and after treatment. Other
pubs shown are the Whittaker Arms, also in
London, and the Druids Head and the
Cross Keys, both at Brighton.

1952
Examples from a corporate design programme
for W. & A. Gilbey Ltd (now International
Distillers & Vintners) which began with the
design of bottles, labels and packs for the
company's wines and spirits and was later
extended to include a complete house style,
shop and office interiors, and consultant
architectural work for the company's head
office in London, and for distillery and
warehouse buildings. *Below, left*, is a view of
the directors' lunch room and bar in the office
penthouse at Harlow. *Bottom, left*, is the
Harlow office block from the north west
with the single storey, flint-faced warehouse
on the left of the picture. Architects: Peter
Falconer & Partners. *Below, right and bottom*
are examples of label designs and specially
designed bottles. In 1965 Gilbeys were
awarded the Royal Society of Arts Presidential
Medal for Design Management.

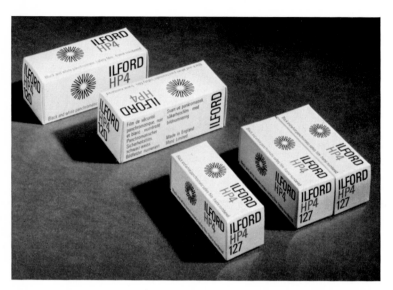

1946–66
The illustrations on these two pages show
examples from an extensive design programme
carried out for Ilford Ltd. This is a second
generation programme to replace the earlier
design scheme which was begun soon after the
war. The new symbol expresses the image of
a camera shutter and the starred effect of a
light source. The camera, *left top*, was an
inexpensive 127 size in moulded grey
polystyrene. The film packs, *second from top*,
show the attention given to clear identification
on all sides. Full instructions for the
implementation of the programme are
contained in a design standards manual, *third
from top*. The Ilfoprint Processor, *bottom*,
received a Council of Industrial Design award
in 1967. *Opposite* is shown the application of
the house style to the transport livery.

1956–66
The design programme for British Rail is the largest and most complex of any attempted in this country. It began in 1956 with the establishment of a BR design panel which has planned and co-ordinated all subsequent work in this field. Many industrial designers were brought in to work with BR's engineering and other staffs. DRU's involvement began with design consultancy on seven locomotives, one of which, the D.1000 series, is shown *left, bottom*. The collaborating designer was J. Beresford Evans. In 1963 DRU developed a corporate identity programme which is shown here applied to the SRN4 Hovercraft, *left top*, freight-liner containers, *left centre*, car ferries, *opposite, top*, and station names and signing, *opposite left*. Extensive studies were carried out into the design of new staff uniforms, *opposite, far right*. Instructions on the application of the scheme are contained in corporate identity manuals, *opposite bottom*. Kinneir Calvert and Associates designed the type-face and collaborated on the scheme for the station sign posting. The centre photograph, *opposite*, shows an experimental sleeper compartment designed for quick conversion to first and second class use.

1968
These illustrations show rolling stock and stations for the new Victoria Line. The visual aspects of the line have been supervised by a design panel, a member of which is Misha Black, London Transport's design consultant. Some detailed design work has been carried out by DRU working in close collaboration with the London Transport Architect, Chief Mechanical Engineer (Railways) and Chief Signals Engineer. The new line continues London Transport's tradition of high design standards which began under Frank Pick's direction after the First World War.

Chapter 4   **Professionalism**

The meaning of professionalism

In previous chapters we have referred to the 'profession' of industrial design. Whether or not industrial design can be regarded as a true professional activity has been a continuing preoccupation among practising designers since before the war. The fact that it should have been the subject of argument at all is a reflection of the uncertainty which many people have felt about the precise nature of the activity which the term 'industrial design' describes. The difficulty is that industrial design does not fit neatly into an easily recognized compartment. It embodies elements from a host of other skills, from engineering, architecture, and painting, the modern skills of industrial management and marketing, the human and physical sciences, technical skills such as photography and dozens of different productive and reproductive processes. If you say 'doctor', 'solicitor', 'journalist', or 'architect', most people have a fairly clear idea of the broad nature of the activities which these terms represent. But if you say 'industrial designer', even after more than twenty years of active promotion, people are still not sure exactly what this means.

Misha Black summed up the dilemma in a talk he gave in 1964 at a meeting of the two leading American design societies, which have since combined into a single body called the American Society of Industrial Designers:

'If industrial design were only a matter of science and technology', he said, 'we could be well content. . . . But we know it is not as simple as that. Art and design are somewhere linked. We know that we are not only engineers concerned with man-machine relationships, ergonomists, marketing experts, and the rest of our public relations image. We really care about what the products we design look like and feel like, and how well they serve the community. We glow with modest pride if one of our designs is acclaimed by other designers as being of outstanding quality. Sales alone (while primary to our very existence as designers) are not the only criteria by which we measure our success or failure. We retain secret standards for our judgement of design merit which sometimes give us momentary pleasure and more often prolonged anguish. Industrial designers do not judge their work only by the standards of their clients' sales graphs and dividends.

This is part of the process which entitles us to be accepted as a profession and not a trade.'

On the face of it, this concern with professionalism could be dismissed as an unimportant matter of terminology with little relevance to industry or to the general public. But already it is apparent that professionalism for the designer is as necessary to his particular occupation as the Hippocratic oath has been to medicine. The attitudes which the term 'professionalism' stands for have certainly influenced the development of industrial design and have helped towards its acceptance by industry and the public.

In 1930 a small group of designers, concerned mostly with the graphic arts, joined together to create the Society of Industrial Artists, the first society of its kind in the world. Prominent among this group was Milner Gray, who remembers 'I was one of a small band of designers and illustrators who felt so passionately about the need to improve the lot of ourselves and our kind that we decided to band together to form a professional association; so eager were we that we achieved our first goal after only eighteen months of argument and debate. . . . At the time of these beginnings of our Society, it was laid upon us by the external authority of the Board of Trade that we should not be permitted to become a trade union, and so it is that, by accident and/or design, we lay claim to being a professional association.'

By the outbreak of war, the SIA had 300 members and had widened its interests to include exhibition and product designers. Its real development, however, began in 1945 when it disbanded all its current members and set up an examining board to vet applications for re-admittance – with the result that some original members failed to get back. Today there are about 3500 fellows, members, licentiates, and associates, covering all aspects of industrial design, but still with a considerable emphasis on graphics. Associate membership was introduced within the last few years to cover people who are concerned with the administration or promotion of design work, and with design education, but who are not practitioners. The Society has also widened its interests to include a sprinkling of engineers, the most notable of whom is Sir Barnes Wallis, the designer of the Wellington bomber, the 'Dam Busters' bomb, and many other well-known engineering projects, including the invention of the swing-wing configuration for supersonic aircraft.

These changes have acknowledged the developments in industrial design which have already been referred to, while more formal recognition was given rather belatedly in 1963 by a new name – the Society of Industrial Artists *and Designers*.

Designers had long appreciated that 'professionalism' was being used by different people to mean entirely different things. To the sportsman it was used to distinguish between a player who earned his living by his sport from one who took part in it as an amateur for his own pleasure. To the craftsman or specialist it was used to denote the degree of skill in workmanship. To the class-conscious, it was used as a mark of superior status within the community. To the professions themselves it was often used to describe a form of behaviour or moral responsibility to society, overriding the opportunity for personal reward. Designers would argue for any one of these interpretations in preference for another, but in fact they are all relevant.

One of the professional bodies to which the SIAD turned, when it set up a committee in 1958 to define the implications of professionalism to the society and its members, was the Institution of Electrical Engineers. W. E. Wickenden, a former president of the IEE, had set out in a paper 'The attributes which mark off the life of a corporate group of persons as "professional in character".' He listed six such attributes, putting first the need to build up a body of knowledge which was common to all members of the group; secondly, the responsibility to pass this on to a new generation through education; thirdly, the need to establish a recognized professional qualification for members; fourthly, a code of conduct; fifthly, the establishment of a recognized status; and finally the creation of an organization to promote the advancement of the other five attributes. The SIAD accepted this definition as a basis upon which to develop its own organization.

For DRU this philosophy of professionalism has had a special importance, partly because it has to some extent controlled the evolution of its own practice and partly because, as a group, the Unit has probably contributed more, over the years, to its realization than any other design office. Individuals in the group have done this by almost continuous participation in the Society's work since its foundation; by teaching and by contributing at a higher level to the direction of design educational policy; by lecturing and writing; and by promoting an international understanding and exchange of ideas. Not every senior member of the Unit, of course, could claim to have contributed in all these ways to the growth of their profession. But the fact that DRU is a large organization with a long history means that almost inevitably its collective contribution adds up to a considerable share of the total effort. By the same token, the fact that the contribution of a single design office can be so great, is a reflection of the size of the society as a whole when compared with other professional bodies of recent origin.

Compared with the SIAD's 3500 members (achieved, it must be emphasized, only as a result of an active promotional campaign) the membership of the Royal Institute of British Architects stands at some 18,000, while membership of the Institute of Mechanical Engineers is around 70,000. The hard truth is that numbers multiplied by subscriptions equals income, and the SIAD has undoubtedly been handicapped by its limited resources in achieving nearly all the attributes of a profession which Mr Wickenden had so ably described.

Income, however, has not been the only factor which has inhibited the creation of a body of knowledge about industrial design. For an activity which has rated such a large degree of promotional interest over the years, knowledge, in the form of accepted laws and principles, is remarkably absent. This is largely due to the fact that industrial design is essentially a cross-disciplinary activity which draws on knowledge from other fields. It is seldom necessary for the designer to be a specialist in any of the skills he draws on – if it were, his training would be so prolonged that he would still be a student in middle age. His own special skill is in adapting the knowledge of others to serve the creative, problem-solving process which is his main preoccupation.

The industrial designer is at a disadvantage here compared, for example, with an engineer. A mass of experimental evidence will enable an engineer to calculate that a beam of given material and dimensions will carry a certain load. But there is no similar information available to the industrial designer which will enable him to calculate that a product, embodying certain characteristics of form and colour, will achieve a predictable result in terms of sales performance, or will make a substantial contribution to the cultural well-being of the community. In this sense, designing for human appeal and satisfaction is as much an art (and thus is as subject to personal whim and fancy) as it has ever been. Where knowledge is beginning to be collected is in the study of the design process itself. We have already described how interest in systematic design methods grew from an earlier awareness of the relevance to design of the human sciences. That the S I A D has done little, as an organization, to encourage interest in this work is due not only to lack of funds for serious research but also to a certain lack of interest in the subject among the majority of its members. But proposals to remedy this situation have been formulated in recent years and we shall refer to these later.

There are a few, of course, who have always recognized that a body of knowledge about the activity of designing is of fundamental importance to the future of the profession. Among these, Misha Black had the rare opportunity, in his capacity as Professor of Industrial Design (engineering) at the Royal College of Art, to provide facilities for a continuing research programme into all aspects of the design process. That he accepted this challenge in spite of personal convictions about the large measure of 'art' that will always be present in the design mix, is characteristic of his open-minded attitude which has been a bulwark against complacency in D R U's own development. As we have previously described, the design laws are beginning to emerge from the R C A research unit, tested against realistic projects which, in themselves, are fascinating studies of innovation. The implications of this work, however, are likely to be understood more by a generation of designers who will emerge in the future than those who are in practice today.

Revolutions in design education

It would be wrong to suggest, of course, that published work on design methodology represents the sum total of knowledge about industrial design. If it did, then the educational process, referred to by Wickenden as a necessary attribute of a profession, could hardly have existed at all for industrial design. In fact, there has been design education in one form or another since the 1830's, following the recommendations of the Ewart Committee, and since the second world war it has been the object of almost continuous and searching debate. That the process of designing for industrial production is subject to procedures of analysis and synthesis as a framework for creative work had been clearly demonstrated in *Britain Can Make It*. It was a framework which could certainly be taught on the basis of knowledge accumulated by practising designers in industry. But as such it was prone to a good deal of personal interpretation, and lacked the more formalized procedures which the later research workers into design methods have attempted to provide.

The first major revolution in design education came in 1948 when the Royal College of Art was reorganized by Robin (now Sir Robin) Darwin following his appointment as principal. The RCA, originally established in the 1830's to train designers for industry, had soon drifted into a school for the fine arts and a variety of craft skills. It set a pattern for all the other art schools which grew up during the nineteenth century, and this pattern persisted for over 100 years. Darwin set out to change this in a manner which was swift, bold, and ruthless. The orientation was switched from arts and crafts to design for industry. A series of specialized design schools were created and, though the fine art departments were retained, a separate diploma for the designers was established. Previous members of staff were sacked, and the shabby, run-down environment, so characteristic of the art school atmosphere, was changed into an efficient working environment, later to be an essential feature of the College's new building opened in 1964.

The revolution at the RCA had a profound impact on design education in Britain. Unlike the college's single-minded progress, however, design education elsewhere has been plagued by uncertainty and vacillation. The National Diploma in Design (NDD) which was introduced after the war to replace the earlier Ministry of Education intermediate examinations, was largely in line with the new trend towards specialization which was the basis of the RCA philosophy. Only by acquiring a high degree of skill in a comparatively narrow field, it was argued, could the student possess the necessary foundation for widening his experience later on. Under the NDD system, however, specialization was carried so far that the art schools were widely criticized for turning out low grade design technicians without the academic background and breadth of experience needed to upgrade design in the scale of industrial priorities.

The situation was ripe for change when, in 1960, twelve years after Darwin's arrival at the RCA, the second revolution in post-war design education took place. The revolution, in the form of a report of the National Advisory Council on Art Education (the Coldstream Council), swung the emphasis right away from specialization towards a more broadly based curriculum as lacking in specialization as the pre-war examinations in art and crafts had been. The cue for the new approach was taken from the universities where the ideal of 'education' in the widest sense was seen in contrast to the 'training' provided by the technical schools. A much higher standard of academic achievement was demanded as a basis for entry to the new courses, to combat the traditional image of the art school as a repository for those too dull-witted to succeed at anything else. A new Diploma in Art and Design (DIP.AD) equivalent to a university first degree was to be introduced. To ensure that the requisite standards could be achieved, the number of schools providing diploma courses was to be severely limited, and a strict vetting procedure was to be introduced to weed out those schools with inadequate facilities. A new National Council for Diplomas in Art and Design (the Summerson Council) was set up to administer these proposals and the first diplomas were awarded in 1966.

The DIP.AD has a special significance to the profession of design for two reasons: first, because the SIAD was actively involved both in the discussions which took place at the advisory council stage, and more especially during the vetting procedures when teams of designers and others visited all the art schools which sought approval for their courses; and secondly, because the DIP.AD provided for the first time a recognized qualification for entry to the profession – the third of Wickenden's attributes.

The need for such an entry qualification had been recognized for many years by members of the profession, and its absence prior to the introduction of the Dip.AD had severely undermined the SIAD's claim of professional status for industrial design. But the controversy over design education was by no means halted when the decision to go ahead with the Dip.AD proposals was finally agreed. Some might argue that it was only at this point that the troubles really began.

The process of implementing the advisory council's proposals created turmoil among the art schools. Those whose courses were recognized found themselves faced with major reorganization. Those whose courses were not, felt that they had been left with nothing to do. There were plenty of doubters about the broadly based approach of the Dip.AD. Even the old NDD courses provided insufficient technical training to satisfy the requirements of industry, and the Dip.AD, it was argued, would simply make matters worse. It would produce a new generation of designers with big ideas, but with so little design expertise that industry would find no use for them, even though certain subjects like interior design and industrial design (engineering) were to have an additional two years built into the courses.

The SIAD came to the rescue with a proposal for a series of courses for the 'have not' art schools which once more resurrected the specialized approach of the NDD and provided a further qualification for entry to the profession. The pendulum, it seemed, was swinging backwards and forwards so fast that it was almost impossible to see it.

In fact these new 'direct entry' courses, as they were called, have provided a genuine alternative to the Dip.AD, and their introduction at least acknowledged that there were 'more ways of killing a cat than by choking it with cream'. But, having settled down to accept this dual system as a pattern for the future, design education received a body blow from an entirely unexpected quarter.

The third revolution, which started about a year before this book was written, came not from Government committees, or industry, or the profession, but from the students themselves – students, it is interesting to note, who had come in through the new, fine-meshed academic sieve which forms an essential part of the Dip.AD system. Although this student revolt in some leading art schools had a variety of causes, a lot of the criticisms centred around the apparent irrelevance of some aspects of the curricula which the broadly based, liberal approach of Dip.AD had tended to encourage. The students were clearly worried about where all this would lead them. And once more it inspired an anxious flurry of rethinking among the educationists. If the eventual outcome is a more realistic study on a national scale of industry's future need for designers, then the student rebellion will have been well worth while. Such a study is certainly needed, for there is plenty of evidence that more designers are now being trained than industry can expect to absorb in the near future.

Design Research Unit's contribution to this educational process is most evident in Misha Black's work at the Royal College of Art. But other partners and associates in DRU have helped to evolve the educational policies we have described, and must therefore share the responsibility for their successes and failures. Black and Gray are both members of the Coldstream Council and they have both been chairman of the SIAD's own education committee, which has been primarily concerned with the relationship between design education and the entry requirements of the profession. A number of other DRU partners and associates have contributed to the work of the Summerson Council and are members of the governing councils of art schools, or give lectures to students. Much of this part-time work is unpaid and reveals a profound interest in problems of education for the profession which is shared by nearly all practising industrial designers.

Misha Black's own philosophy of design education has been evolved during his work at the RCA, which in turn is based on his experience of design in practice. He has established contacts with engineering training centres and has taken a leading part in international discussions, particularly during his chairmanship of ICSID seminars in Bruges, Ulm and Syracuse, organized under the auspices of UNESCO from 1964 to 1968.

His approach to the training of industrial designers for the engineering industries was set out in two lectures he gave at the Royal Society of Arts in 1965. In these lectures he described five basic requirements for the proper function and growth of a school of industrial design (engineering). 'First', he said, 'it required adequate workshop facilities; secondly, it must be closely associated with schools from which it can draw the technological assistance on which it is partially dependent; thirdly, it must be provided with facilities for properly organized general studies; fourthly, its staff must be of a calibre equal to that of the students it aims to educate; and lastly, it should enjoy reasonable academic freedom within the environment of an academic establishment with wider interests than its own.'

But while this describes the mechanics of the educational process, it presupposes a purpose to which it must be directed. The value of industrial design for the engineering industries has often been questioned, particularly for heavy capital equipment where improvements in appearance will not necessarily have a material effect on sales. That industrial design has more far-reaching objectives than these strictly commercial considerations was emphasized later in Black's RSA lectures: 'Its value in improving the function of the artefacts of our civilization requires no advocacy', he said. 'If a hospital bed can be made more efficient, a kettle designed to pour without the lid falling off, if the controls of an electric locomotive be arranged to reduce the danger of accident through fatigue, then industrial design has made a contribution to our well-being. But that concern with comfort and efficiency would not alone justify the high academic and social rating which I claim for this new profession. The real value of industrial design, and the justification for its growth, reside in its being a minor art form. Sir Herbert Read, in his brilliant paper "The Disintegration of Form in Modern Art", separated out two aspects of art ". . . one of a mathematical nature which gives rise to the category of beauty and one of an organic nature which gives rise to the category of vitality. The greatest works of art are those that combine these two elements in a form which we call organic because it possesses both beauty and vitality." He developed this argument by quoting Ruskin, who said that all genuine works of art were the result of "mathematical laws organically transgressed".

'Industrial design, by its very nature, is debarred from the vitality
which is dependent on the uniqueness of a work of art. The fact that
it is art-once-removed, and dependent on exact duplication which is
inimical to personal expression, removes from industrial design the
organic properties which are an essential ingredient of art. But the
mathematical beauty remains in even the most polished machined
object and, if that be only one of the constituents of art, its
multiplication in thousands or millions by mass production gives this
truncated art form a validity which justifies the passionate dedication
of industrial designers to their work.'

Not everybody would agree that art, however minor, is the prerogative
of industrial design within the engineering industries, and such
distinctions may not be altogether helpful to the establishment of a
more unified concept of the design process as a whole. But the
argument against the kind of philosophy set out by Misha Black is
not so much that the idea of industrial design as an art should be
discounted, but that the creative element is essential to all aspects of
the design process, whether it is called engineering, architecture or
industrial design. That concern for human values has often been
missing in engineering design is indicative of the failure to recognize
that creativity often comes from the interaction of all relevant factors.

Organizing for the future

The emphasis that industrial designers place on the wider implications
of their work is part of their claim for professional status. But, like
justice, which must be seen to be done, professionalism must be
embodied in a code of behaviour which binds its members and
proclaims to all the transcendence of social purposes over and above
personal gain. The temptation for a new profession is to acquire the
attributes of its elders, and the SIAD, in attempting to embody the
fourth of Wickenden's principles, has tended to play for professional
safety rather than to give reign to youthful rebellion. Nobody quarrels
with most of the clauses which make up the Society's Code of
Professional Conduct, but a comparatively recent clause clamping
down on the advertising of design services created dissension leading
to resignations and dismissals. While the more vigorous aspects of
this clause have been modified since it was first introduced, the society

Opposite : The members of DRU in 1968.

is once more examining its rules of conduct in the belief that where ethics are concerned there are no absolute values. Speaking of this new study of the SIAD's codes of professional conduct in his Presidential address in 1968, Milner Gray said: 'It may be that we shall evolve new concepts of responsible practice, and devise new codes to guide our members, releasing them perchance from some unnecessary restrictions, as we know that other professions are considering to do. But we shall not forget that the object of all professional codes of conduct is to ensure that everyone who consults or employs a member of such a profession can have trust in him and can rely on his integrity and fair dealing; and equally that he will behave equitably in his relations with his fellow members and honestly with the community as a whole.'

One of the major difficulties with the existing rule on advertising is that a large proportion of the society's members are working directly for the advertising industry, and for most others the commercial interests of their clients must be a major consideration. It may seem strangely illogical for designers to impose on themselves standards of behaviour which are out of key with those of the people for whom they work, for a designer must identify his approach directly with the interests of his client if he is to succeed at all. The architect's professional trust has traditionally been considered as a method of protecting the interests of the client in his relations with the builder. In a similar way it has been argued that the real client of the industrial designer is not his employer but the consumer. If the employer's brief is such that the designer considers that to fulfil it would be to the detriment of the consumer, then he has no alternative but to relinquish the job.

In practice the issue often turns on a question of design standards which, to a large extent, the designer must establish for himself. To prepare designs of a standard which falls below that which the designer considers to be an acceptable minimum, would represent a betrayal of his professional trust, just as an engineer who ignores structural safety limits for the sake of a cost saving would similarly betray his own professional obligations.

This argument, of course, tends to paint a picture of industry as an

ogre, growing fat on profits corruptly obtained from a gullible and
unprotected public. In most cases industry is at least as conscious of
its social responsibilities as the designers which it employs. There are
cases, however, when designers have felt that the demands of their
clients are unacceptable. DRU has experienced such cases and, like
other designers or design groups, has had reluctantly to withdraw
from the commission.

Such attitudes can only emphasize the status of the profession in the
eyes, not only of the public, but of the industry which the profession
serves. They represent a long haul from the days in 1937 when, in its
report on *Design and the Designer in Industry*, the Council for Art and
Industry said of the designer's status: 'Broadly speaking, the evidence
we have received shows that at present industrial art designers have
failed to attain a sufficiently definite and established place in industry.
Too frequently their status is entirely subordinate, and in consequence
they exercise little control over the output of industry.'

Yet, although great progress has undoubtedly been made since then,
there are achievements which still escape the profession. Wickenden,
among his six attributes, referred to the recognition of status being
a matter not only of colleagues and clients, but also of the State. One
form of recognition by the State would be a royal charter, such as has
been accorded to architects, doctors, and others. Social conditions are
changing fast and the validity of royal charters as a form of State
recognition has recently come to be questioned, but the SIAD still
hopes that in the not too distant future it may be given either a royal
charter or its equivalent, since this would bring recognition, too, from
those to whom a chartered official stamp of authority brings a
comforting guarantee of competence and ethics.

But if, until now, the Society has suffered from lack of this sort of State
recognition, it has been equally plagued by the problem of establishing
a form of organization for itself which can cope with its unique
professional difficulties in embracing as 'industrial design' such diverse
activities as book illustration, product design, textile design and the
design of the interiors of buildings, and including both independent
and staff designers.

This diversity within one society makes the problem of organization peculiarly difficult. Although all the individual groups within the society are held together by their common concern for human values in design, they have established for themselves quite different traditions and methods of working. These differences are so great that they make it impossible to establish a code of conduct, a scale of fees, or an educational process, which applies in all respects to all the groups. Free-lance textile designers have a tradition of submitting collections of their work to potential buyers. Yet the code of conduct prohibits designers from advertising their wares in this manner, and an exception to the rule had to be written into the code to take account of this, for otherwise such designers would have been deprived of their living. Similarly, different groups of designers have developed different methods of charging fees. Interior designers, like architects, charge a percentage of the contract price of the job, while product designers tend to base their fees on an hourly rate, and free-lance illustrators or textile designers normally charge a set fee for each drawing or design.

The effect of these variations is to complicate the work of the special committees, since the implications of any decision must be examined for all the different groups. Inevitably there are delays, and a heavy burden is placed on the few public-spirited members who, together with the small permanent staff, carry the weight of the Society's work. Compared with its opposite numbers in countries overseas, the SIAD is exceptionally prosperous, since it is not only the oldest professional association of its kind but also the largest. In spite of this, however, it has always lacked sufficient funds to support its professional work, and, in 1964, it set up a national appeal to British industry with the object of raising half a million pounds to put the society on its feet. The appeal, which was recognized as being charitable, had an impressive list of sponsors, with Lord Chandos as chairman of the appeal council and Milner Gray as chairman of the board of trustees. The purposes of the appeal were 'to establish and endow a building equipped with a library, lecture rooms, exhibition hall, and the necessary offices; a board of design education and facilities for research; and a comprehensive library and information service which will enable those concerned with industrial design to keep level with, and if possible in advance of, design thinking throughout the world'. The programme of research was to include 'product planning and design rationalization; methods of selecting designers for permanent posts; research into the requirements of postgraduate training in industrial design; and an assessment of the country's future requirements for industrial designers in various categories'.

At the time of writing the appeal had raised about £86,000, of which £20,000 were contributed by the society's own members. Although this is a long way short of the target, which means that most of the objectives of the appeal are still as far from realization as they ever were, it is a start, and has put the Society in a stronger position than it has been in the past. Fortunately, another event intervened which allowed the first aim of the appeal, the provision of a new headquarters building, to be achieved far sooner than anybody had expected.

In 1965, a row of elegant Nash houses at the east end of the Mall became vacant, having been used since the beginning of the war as Government offices. Sir Roland Penrose, chairman of the Institute of Contemporary Arts, had the idea of inviting a number of societies to join with the ICA in taking over part of the terrace and creating a headquarters for a new Association of Societies of Art and Design. In addition to the ICA itself, there were the SIAD, the Design and Industries Association, the Designers and Art Directors Association, and the Institute of Landscape Architects, all of whom similarly lacked adequate headquarters facilities. They were enthusiastic about the idea, and the Carlton House Terrace project, as it came to be called, went ahead with Sir Duncan Oppenheim as chairman of the board of management. An exhibition hall, conference suite, library, restaurant, committee rooms, and offices for the individual societies were planned, and for the first time the prospect of a permanent home for the SIAD, with all the facilities it had ever dreamed of, could at last become a reality.

It is expected that the conversion work will be completed by the summer of 1969 when one of the basic tools will have been forged with which a form of organization appropriate to the needs of a professional body can be built up. This had been the last of the attributes of a profession which Wickenden had set out in his paper. But the existence of a satisfactory headquarters building does not itself solve all the problems. The small size of the Society remains as the basic weakness in the development of a form of organization capable of meeting the profession's future needs.

In recent years the SIAD has failed to capture the interest of some of the more creative and successful of the younger designers, who tend to regard its activities as a sap to the vitality which they see as central to their work. The move to Carlton House Terrace is likely to confirm rather than overcome these attitudes because, even within the ranks of the society, there are plenty who regard the move as an expensive and irrelevant exercise in pomp and circumstance. But even if all these dissenters were brought into the fold, the effect on the SIAD's size would not be of great significance. Steady growth is likely to come about more from the art school graduates, whose diplomas now provide the main form of entry to the profession. Even here, however, dramatic growth cannot be expected for, in the future, the effect of mergers in industry and the design rationalization programmes which result from them, is likely to dampen the growth in the number of designers which industry requires, and this in turn will affect the number of industrial designers being trained.

In these circumstances, the SIAD must either reconcile itself to being a small professional group, with all the limitations on its activities which this implies, or it must seek entirely new ways of overcoming the problem. In the next chapter we will suggest that changing attitudes during the last decade will enable a new philosophy to be produced which is capable of embracing a wide range of specialized activities all concerned with the creative activity of designing for industrial production. The creation of a professional association to represent designers in all fields might seem like a remote pipe dream to some, and to others a betrayal of everything that had been achieved in the battle to gain recognition for industrial design in its own right. But, as we shall see, new developments which are now beginning to take place may well lead to a closer integration between engineering and industrial design. If this happens, as it surely must, then the climate will be there for some kind of merger which would bring industrial design, engineering, and perhaps architecture together into a single design profession of unprecedented power and influence. Such a professional group will become increasingly necessary if the design challenge of the twenty-first century is adequately to be met.

Chapter 5   **Future**

A new philosophy

In the previous chapters we have discussed some of the major changes in the theory and practice of design for industry which have taken place since the beginning of the industrial revolution. We have seen how a major split developed during the nineteenth century between what we have described as the art-oriented and science-oriented design streams, and how, almost a century later, interest in the human sciences, together with the evolution of design services to meet particular problems in industry, have laid the foundations for a closer understanding between the designer's ideals and the practical needs of industrial management.

Perhaps the most important outcome of these new attitudes is that they have opened up the way to a design theory which extends far beyond the boundaries of industrial design, and is capable of embracing products or design programmes ranging in size and complexity from egg cups to aircraft, and from sinks to cities. Moreover it enables the traditional skills of the industrial designer to be seen in a proper relationship with the design process as a whole, depending on the type of product or design activity involved.

Developments in production technology since the war have become so advanced that the requirements of production impose fewer constraints on the designer than they did in the past. In the early post-war years it was frequently argued that the aesthetic of the machine age derived directly from the production techniques used. It was suggested that you could tell if a form was good or not simply by assessing whether it conformed to good practice in moulding, pressing, casting, or whatever technique was being used. But only the expert, of course, was able to discern these qualities. The consumer was seldom aware of them, except where the quality of finish was so poor that it became a physical or visual irritation.

It was suggested that a plastics moulding, for example, should be slightly tapered, with rounded corners to enable the product to come easily out of the mould. In the same way, it was also argued that a refrigerator door had to have a deeply curved form because this suited the metal pressing process, and pressed metal was the only technique which could provide a door that would be stiff enough to ensure a proper seal against the cabinet. But hardly had these ideas been propounded, than perfectly finished plastic mouldings and thin, rigid refrigerator doors were produced in severe, sharp-cornered, rectangular forms, confounding the theorists. What had happened, in fact, was that, for several reasons, the requirement had changed and the production engineers had quickly devised methods of meeting it.

As far as industrial design was concerned, the emphasis had again shifted. Milner Gray, in his RSA paper of 1949, had described how the emphasis had changed from 'design for selling' before the war to 'design for making' immediately after it. Now it could be said to have moved to 'design for consumption'. That consumption was the sole purpose of all industry was the battle cry of the consumer protection bodies, and it was to be increasingly accepted by industry through its growing emphasis on marketing.

The pressure of all these events has caused the activity of industrial design to be associated more directly with those factors which are concerned with a product's use, and less with those that are concerned with its production. These human, user factors embrace such product qualities as handleability, ease of operation, comfort, ease of cleaning and maintenance, shape, weight, colour, and so on, and have a proportional relationship with the technical requirements which varies according to the type of product. The design of a printed curtain material is almost entirely concerned with human factors and is wholly the job of the industrial designer. A jet engine for an aircraft, on the other hand, is almost wholly concerned with technical factors and the industrial designer has no place in its design. In between these extremes are a whole range of products in which the ratio of human to technical factors can be roughly determined, and the job of the industrial designer defined. It will differ in almost every case.

The tendency to generalize about industrial design in early writing on the subject failed adequately to take this diversity into account. The result was a growing conformity of style – a notable characteristic of consumer product design in the late 'fifties and early 'sixties. To the leaders of design opinion, there was little room at that time for individual expression. Reaction to the exuberance of the Festival took the form of a severe, angular appearance which expressed a rediscovery of the Bauhaus and the work of the earlier constructivists. But it was a blinkered and inward-looking development which failed to satisfy the emotional needs of wide sections of the general public, and particularly the younger generation.

The Pop movement blazed into this sterile world and started a fire which raced ahead, creating havoc and confusion among the complacent design establishment. It may have succeeded in banishing for a long time to come the idea of any single 'correct' approach to the aesthetics of design.

Stephen Spender recognized this dissatisfaction as long ago as 1958 when, in his SIAD Design Oration, he said: 'In some cases – as in the design of wireless or television sets – the aim seems to be to look functional rather, perhaps, than to be it. The concept of function translates itself into bareness, simplicity, squareness or roundness, solidity, seriousness. Above all, everything is impersonal. The simple furniture, unobtrusive yet tasteful hangings, clean-limbed flat irons, innocuous cups and saucers, limpid glass, governessy electric clocks, glacial refrigerators, flat-chested cupboards, disinfected panels of linoleum, all suggest the young married couple wearing plastic overalls and washing up together after a wholesome and simple meal of foods which are a harmony of oatmeal and pastel shades, at the end of a day in separate offices which are equipped with electric typewriters.'

His words anticipated the revolution that was to come, but it was a lone voice in the design profession that cried unheeded for five or six years.

The lesson the Pop movement has taught is not that the austere styles were wrong, but that different styles can co-exist and be equally good or bad as the case may be. It established, too, that aesthetics in design are not something which must be slipped in unobserved under a strict cloak of practicality and common sense, but a fundamental human need which is now recognized by designers, manufacturers, and public alike. The movement had started with a 'Teddy boy' taste for Edwardian clothes, later to branch out in the more varied collection of 'mod' outfits. It was eventually accepted, in a multitude of variations, by all classes of society when its influence spread outside the clothing industry into furnishings, interior design, advertising, and almost any situation where the ephemeral nature of the product could justify delight in extravagance.

Artificial obsolescence, so roundly condemned by designers in the 'fifties as a threat to the permanence of their work, was now eagerly sought as a positive enrichment of life. The desire for change was firmly established as an inescapable factor in the design of many products. Seen simply as a measure of the rate at which change takes place, fashion could be accepted as having a legitimate place in the philosophy of design. With the breakdown of belief in absolute values, to describe a product as 'fashionable' no longer implied condemnation as it had done a few years before. The rate of change simply varies from one type of product to another. In clothes it may be six months, in furniture ten or twenty years, in houses 100 years. It would be as unreasonable to make a dress so strongly that it would last for a decade as it would to make a house so flimsily that it would constantly need replacement; for, while people desire change in some things, they equally desire a sense of permanence and continuity in others. The advocates of disposable houses might well find their market as small as that of the makers of long-wearing dresses.

What is clear from all this is the discovery that, just as the ergonomists had first shown, neither 'the average man' nor 'the ideal man' exists in sufficient quantities to make it worth while bothering with them. This is as true for aesthetic and emotional needs as it is for physical ones. That it poses difficult problems for an industry conditioned to the idea of long production runs of precisely the same article is undeniable. Already, however, the more advanced industries have provided a clue to the way in which a variety of individual needs can be reconciled with production in quantity. In the motor industry a large number of variations on a standard model are possible, so that when he orders a new car, the consumer's particular choice of features (colour, engine size, extras, number of doors, and so on) can be recorded on a sheet and fed into a computer which orders the necessary parts at the right time and place on the production line. The result goes some way towards the idea of a tailor-made product designed to the consumer's individual needs.

Extension of such a process to other industries will become possible when production is increasingly geared to the sophisticated computers of the future.

Industrial design and engineering

The connexion between Carnaby Street and the engineering industries may not be obvious, but design attitudes in both stem from the more systematic study of the design process which has taken place during the last ten years. The failure in engineering design is that the design activity itself has got lost among the ever-increasing flood of technical information which the engineer must digest. In a good deal of engineering design the process is complicated by the fact that design teams are often considerably larger and more varied in the specialized skills involved than they are in the consumer products field. The possibility of obtaining a well-integrated result is therefore more remote unless there is a strong lead from the head of the team. Only recently has the design process been studied in the training of engineers, and it has been estimated that a graduate engineer will need a period in industry of up to ten years before his competence as a designer can be fully established through a process of practical experience.

The lead given by the art schools in providing practical training in the process of creative designing has already been acknowledged in some engineering training circles. A new emphasis to the importance of the design activity in the training of engineers was given in 1966 by the establishment of the Engineering Design Centre at Loughborough University. Here postgraduate engineers learn to identify the objectives of a design problem and work out solutions, taking into account all the factors involved, including ergonomics and aesthetics. It has been suggested that engineering designers trained in this way would obviate the need for an industrial designer as part of the team.

It has certainly been shown, in short industrial design courses carried out within individual engineering companies, that young engineers and draughtsmen can quickly assimilate the attitudes and skills which the industrial designer can contribute. The more engineers there are who are capable of such work, the better, for the result can only contribute to an increase in the quantity of well-designed engineering products. But it would be nonsense at this stage to put all the eggs in the same basket. Just as the single-minded approach to aesthetics in design proved to be inadequate, so also is a single approach to the training of engineering designers likely to lead to sterility.

As Misha Black said at a conference on 'Industrial Design and the Engineering Industries' at Birmingham in 1959: ' . . . the industrial designer, by training, concentration of interest and personality, differs in important aspects from the engineer. If the industrial designer is properly assimilated within the design team, then these very differences become the flint against which the steel of the engineer can strike more imaginative blows.'

In the past, over-generalization has again led to unnecessary argument about whether or not the industrial designer should be part of the engineering design team. In fact this should depend on the importance of human values in the particular product concerned. To use a medical analogy, if you have a headache you can probably cure it yourself, by taking a couple of aspirins; but if you break a leg it would be wise to go to a doctor. Similarly, simple ergonomic problems can often be handled by an industrial designer or engineer by reference to appropriate data in the text-books, but if it is a complex or unfamiliar problem, requiring fresh experimental research, then the sensible approach would be to go to a trained ergonomist. In the same way a person with four or five years' training in industrial design is likely to make a better job of the appearance and user factors in a product where these are important to its success. Where they are not, then designers skilled in other specialized areas should be able to deal, themselves, with the minor headaches of human values.

What emerges from this proposition is the concept of a body of professional people held together by the common discipline of design; but they will be designers working in a variety of specialized fields, some concerned with structures, some with plastics, some with circuity, some with ergonomics, some with graphics, and so on. The degree of specialization will vary and there will certainly be a need, on the one hand, for highly specialized designers and, on the other, for co-ordinating designers of wide experience. Within this context the term 'industrial designer', referring as it does to only one group of specialists, is highly misleading, for all the specialized designers we have referred to are working for industry. We have explained in the first chapter the peculiar origins of the term 'industrial design'; after fifty years of use and twenty-five years of Government promotion of the activity it represents, it will be difficult to get rid of. Yet it may have done immeasurable harm to the more rapid acceptance of a unified philosophy of design.

It would be far better if some other term, which frankly acknowledged the industrial designer's preoccupation with aesthetics and human values, could be found. Already designers in fields such as textiles, furniture, graphics, lighting equipment, and so on have dropped the prefix 'industrial', and only where the industrial designer becomes a specialist in a team of other specialists does the problem of terminology really arise. In such cases, instead of calling him, or her, an industrial designer (engineering) it might be better to refer to him by the particular aspect of design with which he is concerned – namely, human values. Why not then a 'human values designer'? It should not be too difficult to agree on such a name and gradually to drop 'industrial designer' from the vocabulary. It would be an important step towards a more integrated approach to design in engineering.

The need for such integration has already been recognized at a national level. It took a Government committee under G. B. R. Fielden, which issued a report in 1964, to awaken interest in engineering design. The existence since the war of a Government-sponsored organization to promote industrial design had not failed to strike some engineers as, in itself, a curious phenomenon. While the CoID stuck to consumer goods, it was all quite understandable, but its increasing activity in the engineering industries had drawn attention to the oddness of Government money being spent on the promotion of one comparatively minor aspect of design in engineering, while all the more important aspects were apparently being ignored. The demand began to grow for a new council to do for engineering design what the CoID had so successfully achieved in, for want of a better term, industrial design. The CoID had already recognized, however, the impossibility of isolating the industrial design activity from all the others in the process of engineering design. To bring both sides together, to promote the junction of the art- and science-oriented streams, was seen, in the last few years, as an increasingly important objective. When the idea of setting up a separate council of engineering design was first mooted in 1967, it seemed as though the chance of achieving this objective would be pushed further away than ever.

But, after a good deal of discussion among the organizations concerned, a working party set up by the Institution of Mechanical Engineers at the request of the Council of Engineering Institutions, submitted a report to the Ministry of Technology in September 1968 advocating a single design council to promote design across the whole span of British industry. The report recommended that the new body should be built upon the existing organization of the CoID, and that additional Government money should be made available to cover the new council's work in the engineering field. Thus, the first concrete proposals were made to bring together under one roof the promotion of the two essentially related aspects of design for industry which had been separated for 150 years.

By the time this book went to press, the Government had responded favourably in principle to the recommendations of the working party, and already detailed plans to support the proposals are being prepared by the organisations concerned. If and when the new council is set up, it will certainly be the most important development in design for industry which has come about since the early days of the industrial revolution.

Widening the design brief

What will be the outcome in the future of such a unified approach to design? And how will it affect the solutions of the new design problems which will emerge?

In a talk on the Bauhaus and its influence on design teaching, given in 1968 on Radio 3, Reyner Banham described an exercise set for graphic design students at the University of Southern Illinois. The task given to the students was to design a scheme to publicize a local charitable appeal. Dr Banham described how one girl student presented herself for the final jury with a card index and a telephone directory, but no conventional graphic design material at all. Her argument was that it was a small town where she knew practically everybody, and she could ring them all up. The instructor concerned said he could not accept this as a design solution but, in fact, Dr Banham contended, it could very reasonably be held to be a proper solution to the design problem which had been set.

The story underlines the dangers for designers of accepting a conventional drawing-board role as the one and only approach to their work. There are now plenty of signs to suggest that, in the future, the boundaries set by convention to define industrial design, engineering design, or any other form of design, are increasingly being transgressed so that the edges are becoming more and more blurred. Problems of an entirely new kind are beginning to open up which designers will need to tackle in ways that will demand a new level of resourcefulness and skill.

It would be presumptuous to forecast exactly how design for industry will develop in the years to come. There are, however, certain trends which have emerged during our discussions which are especially relevant as pointers to the future. The beginnings of a new under-standing between industrial design and engineering show that there is now a more widespread acceptance that human and social values in design are of increasing importance in an age which is dependent on rapid change and growth in technology. At the same time these human and social values are in themselves becoming increasingly capable of objective analysis. And alongside these changes has been the development of a common framework of understanding in which specialized design interests can fit together within an overall pattern like the pieces of a jig-saw.

The effect of these changes is to open up prospects for the future which are as exciting in the opportunities they provide, as they are daunting in the immense difficulties which, almost for the first time, they reveal.

Some of these difficulties were referred to by John Christopher Jones in a Third Programme talk in 1966 (later published in *Design* magazine in a revised form). Many products, he said, which have come about as a result of recent developments in technology, have resulted in totally unforeseen side-effects which have been harmful to society. He cited the case of the motor-car, commercially successful in its own right, as an example among a number of others of a new product which has created far more problems than it has solved – causing congestion, delays, and a growing number of deaths and injuries. More recently the series of accidents on British Rail's automatic level crossings has pointed to the failure to take sufficiently into account human or mechanical failures in a system which is perfectly adequate provided that everything goes according to plan.

The implication for designers of these failures is that the design activity itself will be extended to embrace the conscious examination of factors which in the past have often been dealt with by rule of thumb, or have been ignored altogether. Products like cars, cookers, business equipment, and capital goods of many kinds can no longer be considered as isolated design problems outside the systems of which they form a part. The car is part of a system of individual transportation offering, on the face of it, a high degree of individual choice of route and a large measure of personal control over things like speed and where and when to stop. But the development of one group of components within the system (the vehicles) has outstripped development of almost all the others (the roads, signposting, and parking facilities). So the effectiveness of the system is reduced by the incompatibility of the parts which it contains. The result is that cars have been developed to give speeds in excess of 100 mph at the time when speed limits have been introduced to combat the rising toll of deaths and injuries, and when congestion in cities has reduced achievable average speeds to 10 mph or less – hardly more than the horse and carriage transport of the past.

A similar situation exists in the building industry. Peter Jay, writing in the *Architectural Review*[8] about the new problems which have come about as a result of modern building technology says: 'We can only make such buildings habitable (that is, modern buildings of concrete

[8] *Architectural Review*
February 1968

frame construction with light cladding and large areas of glazing) by installing more and more elaborate engineering systems to provide the self-regulating tendency which the fabric itself used to provide automatically, and we are becoming increasingly worried about the environment because the engineering does not always work very well. It is not that we are more concerned about the environment than our predecessors, it is simply that we understand less how well to control it, and the more we struggle to redress the imbalance of our buildings the more obvious the gaps in our knowledge become.'

Mr Jay was describing how the use of modern technology to solve certain problems in building design (the enclosure of the most space at the least cost) had led to unexpected consequences (the creation of extremely uncomfortable conditions for the people inhabiting the building). The need to think increasingly in terms of the design of systems, rather than of individual products or components, is clear from these examples. But however challenging this concept is in itself, it immediately raises further problems of even greater complexity. A building system or a road system does not exist in isolation. It is when one system comes into contact with another that the problems really begin. The introduction of automatic level crossing barriers is an example of what may be described as an interface between the road system and the rail system. The accidents at these crossings suggest a failure to resolve the conflict between the interests of one system (in this case the need to cut railway staff costs) as against those of the other (the need to build in an adequate safety margin for road users without causing them unnecessary delay). Where the automatic crossings fell short of requirements was the failure at the pre-design stage to identify all the factors that would affect the way in which the crossing would be used.

Even greater conflicts arise when the interests of the road transport system come face to face with the interests of people who wish to live in a congenial urban environment. London is an example of a city which has failed to cope with the phenomenal growth of motor transport; Los Angeles an example of a city whose growth has been dominated by motor transport. Neither has solved the problem of achieving an acceptable interface between the convenience of road users and the convenience of city dwellers. On a domestic scale, the design of houses has failed to take account of the increase in the number of appliances, while the provision of more and more elaborate facilities for cooking has strangely ignored the growth of convenience foods.

If this type of thinking is carried to a logical conclusion, then any comparatively minor design problem can soon swell into an issue of world-wide proportions – with the same disastrous consequences experienced by the old lady who swallowed the fly. A number of designers and research workers, however, have suggested that it is possible to cope with the outwardly spiralling consequences of individual design decisions by planning new developments on a systems design basis. A piece of furniture, for example, may in itself be a system of drawers, shelves, and cupboards, but it is also a sub-system of a house, which is a sub-system of a town, which is a sub-system of a region, and so on.

This approach provides a framework for what otherwise might seem to be a huge, amorphous and insoluble problem. It suggests a way out of what John Christopher Jones has described as 'the problem of organizational inertia' which, he goes on to say, 'is clearly a big restraint on our corporate ability to move with the times'. The inevitable conclusion from these developments is that the 'software' stage in designing – thinking, data collection, analysing, organizing – will become more and more important in the future, while the 'hardware' stage – the transformation of data into design – will occupy a smaller proportion of the total design effort.

The design process is already moving in this direction. We have described how, in the field of corporate identity, the amount of planning and organization required before design work can begin has enormously increased in recent years. A similar trend has been taking place in architecture, especially for buildings like hospitals where computers are beginning to be used to analyse the vast quantities of data needed to solve the complex problems involved. But one project in America has taken this a stage further to a point where the designers were concerned exclusively with the software problems, and were able to pass the design stage on to those with more specialized technical knowledge. The project, known as the School Construction System Development, was for a number of schools requiring a high degree of flexibility in use, which was commissioned by a group of Californian county authorities. The designers of this project did not set out to design the building system in a conventional sense, but spent their entire effort on working out in detail all the requirements of use and of the components which would be needed. It was left to the contractor's own technical experts to translate the detailed specification into a design, carrying out the necessary research and development work which the design team were in no position to do themselves. This approach had the advantage of allowing the designers to concentrate on the functional and human problems which they were best qualified to do, while the contractors were, in turn, able to contribute the technical expertise which they possessed to a far greater extent than the design team.

Developments of this kind provide convincing signposts to the way in which designing for the complex systems problems of the future will have to be tackled. Today, however, the development of knowledge about the human and social requirements is at best inadequate and often misleading. As a consequence, these requirements tend to be undervalued in the scale of priorities, for it is understandable that the excitement of technical innovation can sweep all before it, often regardless of whether the result is socially desirable or not.

It may seem strange that it is the industrial designers and architects, who have grown up within the stream of thought which we have described as art-oriented, who have most eagerly grasped at the straws of human and social science. But their idealism has always caused them to reach out far beyond what they know can currently be achieved. Their concern to satisfy aesthetic and emotional needs has led quite naturally to a search for any means of illuminating what is likely to remain a highly intangible subject. In pursuing this aim they have certainly brought home to many engineers and others engaged in industry that the human and social implications are of overriding importance in the head-long race of technology.

Thus, in the future will emerge the need for a new type of designer who is capable of using all the resources of science and technology to conceive and plan a sense of order out of what appears to be an increasing chaos of unrelated problems. That he must be a creative artist is certain, for he must be sensitive to the often inexplicable irrationality and individuality of human need. His palette will be the knowledge that science can provide, his tools the computer and a variety of new techniques in organization and management. The particular brand of expertise which such designers will require is not beyond present-day capability. The difficulty is in establishing the opportunities and the organizational facilities for the large-scale experiments which will be needed to validate systems concepts before they are introduced. Systems in the past have arisen haphazardly as particular needs have emerged and been fulfilled. The growth of populations, the speed of industrialization, the drain on the world's natural resources, will make such haphazard development less and less tolerable in the future.

Through all these changes, the designing mind must increasingly be evident and must have sufficient authority not simply to combat sectional interests, but more to enthuse them with a sense of idealism and shared purpose, as the designers at the *Festival of Britain* had so admirably succeeded in doing.

In putting forward this concept of a designer capable of fulfilling a far wider brief than has ever existed before, we are not suggesting that designers specializing in narrower fields will not also be needed. They will certainly continue to be needed at every level. The point we would make is that designers capable of spanning the spectrum of human and technical factors, within the context of major systems design problems, are missing from the present-day scene. The challenge to all the design professions is to recognize the need, to promote its acceptance, and to make sure that it can be fulfilled when the opportunities arise.

Index

Photographs